Boron Nitride Nanomaterials

Boron Nitride Nanomaterials

Properties, Fabrication, and Applications

Ben McLean
Alister J. Page

Jenny Stanford
Publishing

Published by

Jenny Stanford Publishing Pte. Ltd.
101 Thomson Road
#06-01, United Square
Singapore 307591

Email: editorial@jennystanford.com
Web: www.jennystanford.com

British Library Cataloguing-in-Publication Data
A catalogue record for this book is available from the British Library.

Boron Nitride Nanomaterials: Properties, Fabrication, and Applications

ISBN 978-981-4968-23-2 (Hardcover)
ISBN 978-1-003-31448-6 (eBook)

To Caitlan, Mardi, Sidney and Molly

Contents

Preface

In this book, we establish and comprehensively review the history of boron nitride (BN) materials cubic boron nitride (c-BN), hexagonal boron nitride (h-BN) and boron nitride nanotubes (BNNTs) with respect to their properties, fabrication and application. From the first crude fabrication of each material to the optimization of complex synthesis methods through the exploration of the synthetic parameter space, the field of BN materials has progressed significantly in the last few decades. Thanks in part to highly precise experimental techniques, the properties of each material have been elucidated with the assistance of theoretical computations, such that the atomic contributors to these properties and the routes of enhancing them are apparent. The synthesis of each BN material has progressed markedly to thc points where c-BN is produced globally as a leading abrasive and wafer-scale single crystal h-BN is set to be a part of next-generation electronics as a dielectric. Despite the great research effort devoted to these materials, however, a complete understanding of their properties and behaviour has not yet been realized. Compared to their carbon analogues, there is also a clear lack of theoretical understanding regarding their nucleation and growth mechanisms. These areas are key for the optimization of the synthesis and application, particularly of low-dimensional h-BN and BNNT materials. This book provides the reader with a detailed history of the materials and an insight into the areas that require attention, guiding both experimental and theoretical work in the future across all areas of properties, fabrication and applications.

Since the first synthesis of c-BN under extreme conditions, the fabrication of this superhard material has progressed in terms of finesse and scale. We describe how high-quality c-BN is achieved using chemical vapor deposition (CVD) techniques with desirable grain sizes to optimize mechanical properties.

Industrially, the production of c-BN continues to approach that of synthetic diamond. Increasing the effort toward understanding the mechanism behind controlling the size and directionality of c-BN grains during growth, a material truly rivalling or surpassing the properties of diamond may be achievable. Further knowledge of how c-BN may be doped and how this influences its properties may lead to applications outside of those focused on its mechanical strength.

We show that h-BN, currently the "hottest" BN material attracting research interest due to high-impact prospective applications, has evolved remarkably from an unwanted phase during c-BN synthesis to a key part of future optoelectronic technologies. Advanced experimental techniques elucidate the attractive properties of h-BN. However, there is still a lack of understanding of the interfacial structure and properties of h-BN at a metal surface or on weakly interacting substrates such as graphene. Considering these arrangements are present when h-BN is used in application, it is critical that a combined experimental and theoretical effort is made to enhance the understanding of these interfaces. Currently, h-BN can be synthesized at the wafer scale and consists of a seamlessly stitched single crystal. With further study and optimization, the growth mechanism can be understood and subsequently scaled up to satisfy future dielectric applications, which show incredible promise in the small-scale test devices that we will describe here.

The youngest of the three materials focused on in this text, BNNTs have still grown as a scientific field on their own significantly in the 25 years since synthesis. Highly attractive properties have been revealed for this material in one dimension. However, the best way to utilize these mechanical, thermal and insulating properties is not obvious. While h-BN may see widespread use as a dielectric in stacked heterostructures with graphene, the combination of BNNTs and their carbon counterpart has only recently been achieved, with the properties and potential applications of which essentially unknown. Currently, there has been relatively little theoretical investigation into understanding how BNNTs are grown and how their synthesis may be optimized, despite BOCVD proving to be an effective synthesis route. Scaled-up synthesis of these materials does not seem feasible currently compared to CNTs. Much like c-BN and h-BN, synergic efforts of theory and experiment will greatly benefit the field to propel this material into the forefront of low-dimensional science.

Chapter 1

Introduction to Boron Nitride Nanomaterials

1.1 Introduction

Boron nitride (BN), in its various allotropes and low-dimensional nanostructures, has the potential to rise to the forefront of materials science. By simply changing the arrangement of the alternating B and N atoms, distinct crystalline phases with varied, yet remarkable properties emerge. This chapter introduces BN nanomaterials of all dimensions and their structural properties that arise from tuning bond hybridization and atom packing.

BN was historically considered a synthetic compound after its synthesis in 1842 involving the reaction of molten boric acid (H_3BO_3) with potassium cyanide (KCN)[1, 2] until its recent (2014) discovery in the Earth's mantle.[3] The mineral discovered, Qingsongite is a natural analog of the 3D, cubic phase of BN, cubic boron nitride (c-BN). In turn, c-BN is the structural analog of the most well-known carbon allotrope, diamond. The B-N bond is isoelectronic to the C-C bond, so there are many similarities between BN and carbon structures. The C-C bond is a common covalent bond, each atom contributing one electron for the bond. The B atom in comparison has an empty p orbital and accepts the lone pair of electrons on the N atom to form a coordinate covalent bond with a significant ionic character.

Boron Nitride Nanomaterials: Properties, Fabrication, and Applications
Ben McLean and Alister J. Page

ISBN 978-981-4968-23-2 (Hardcover), 978-1-003-31448-6 (eBook)
www.jennystanford.com

c-BN is built from sp^3 hybridized B and N atoms, as is wurtzite boron nitride (w-BN), an analog of Lonsdaleite.[4] Hexagonal boron nitride (h-BN) is an sp^2 hybridized analog of AA-stacked graphite. Early electron energy loss spectroscopy (EELS) experiments showed characteristic π^* and σ^* energy loss peaks at the B-*K* and N-*K* edges, indicative of sp^2 hybridized bonds.[5] Arenal *et al.*[6] subsequently showed zero π^* resonance of the B-*K* and N-*K* ionization edges using EELS. Rhombohedral boron nitride (r-BN) is less common but analogous to ABC-stacked graphite. The crystal structures of these phases are presented in Fig. 1.1. A single layer of h-BN provides the BN analog of graphene, and rolling this layer into a cylinder gives the analog of a carbon nanotube (CNT), a boron nitride nanotube (BNNT). Other forms of BN have been seen experimentally, such as non-crystalline, amorphous boron nitride (a-BN),[7] and a layered turbostratic boron nitride (t-BN),[8] and a form not yet realized experimentally, a metastable rocksalt BN expected at a pressure above 1,000 GPa.[9,10] This chapter will present the structural properties of each BN phase, with those of each phase presented in Table 1.1.

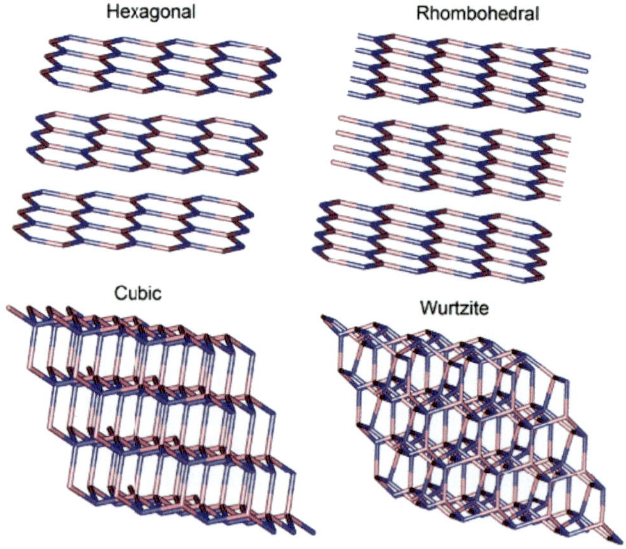

Figure 1.1 Crystal structures of hexagonal, rhombohedral, cubic, and wurtzite BN.[11] Reprinted with permission from Pakdel, A. *et al.* Nano boron nitride flatland. *Chem. Soc. Rev.* 2014, 43 (3), 934–959. Copyright 2014 American Chemical Society.

Table 1.1 Structural properties of BN phases

Phase	c-BN	w-BN	h-BN	r-BN
B-N Bond Length	1.57	1.55	1.45	1.45
Bond Hybridization	sp^3	sp^3	sp^2	sp^2
a (Å)	3.6169[12]	2.5494	2.5043	2.5042
b (Å)	—	4.21[12]	6.6562[13]	9.99[13]
Space Group	$F\bar{4}3m$	$P6_3mc$	$P6_3/mmc$	R3m
Stacking Sequence	ABC...	AA'...	AA'...	ABC...
Density (g cm^{-3})	3484.7[12]	3.467.4[12]	2.2[14]	2.26[15]
Molar Volume (cm^3 mol^{-1})	7.1183[16]	7.145[16]	10.892[16]	10.904[16]

1.2 Cubic Boron Nitride

c-BN was first synthesized in 1957 after subjecting h-BN to high temperature and pressure.[17] The c-BN produced, dubbed "Borazon" by Wentorf and colleagues, exhibited a zinc blende lattice structure and was hard enough to scratch diamond. Subsequent synthesis demonstrated that direct and catalyst-assisted transformation of h-BN to c-BN was achievable.[18, 19] Structurally analogous to diamond, c-BN is chemically inert, electrically insulating, and thermally conductive, having high resistance to oxidation at elevated temperatures.[20] These physical properties are shared between c-BN and w-BN, with the primary difference between them being the internal stacking angles hence lattice structure. w-BN is a structural analog of Lonsdaleite, though is less stable at ambient conditions than c-BN.[4]

1.3 Hexagonal Boron Nitride

h-BN consists of a lattice of stacked layers of hexagonal BN ring networks, structurally analogous to graphite. The stacked layers adopt an arrangement such that B and N atoms of adjacent layers are bonded electrostatically, resulting in the hexagonal rings of stacked layers overlapping one another (i.e., AA-stacking). Conversely, r-BN is stacked in a threefold sequence such that adjacent hexagons do not overlap (Fig. 1.1), akin to AB-stacking in graphite.

h-BN was initially thought to be more thermodynamically stable than c-BN. Numerous studies revised the phase boundary between the two BN phases[18, 21–23] as the high temperature, high-pressure conversion of h-BN to c-BN was the most popular method of c-BN synthesis since its first synthesis in the late 1980s as discussed in Section 3.2.1.[20] In 1986, c-BN was shown to be stable under ambient conditions by Sokolowska *et al.* and with key experimental results by Will *et al.*,[24] the phase diagram proposed in 1999 by Solozhenko is now widely accepted (Fig. 1.2).[25]

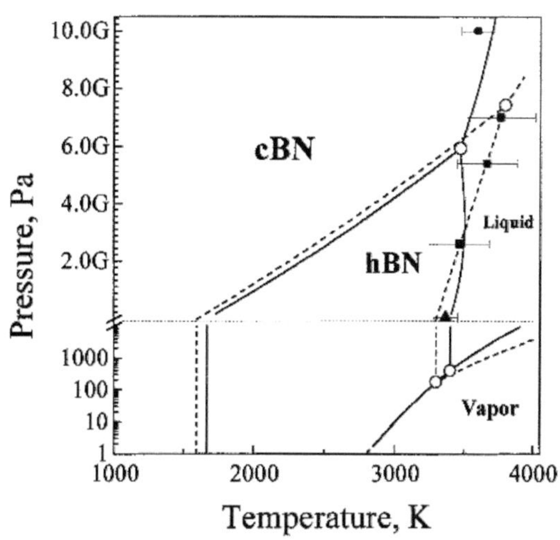

Figure 1.2 Phase diagram of BN. Solid lines compared to the dashed lines indicate the phase boundaries. Solid squares represent h-BN melting, solid circles represent c-BN melting at 10 GPa, and solid triangle shows the melting point of h-BN at 50 MPa.[25] Reprinted with permission from Solozhenko, V.L. *et al.* Refined Phase Diagram of BN. *J. Phys. Chem. B.* 1999, **103** (15), 2903–2905. Copyright 1999 American Chemical Society.

When isolated, a single layer of h-BN exhibits enhanced chemical, thermal, and optical properties, as do single-graphene layers. The edges of an h-BN nanosheet are either zigzag (ZZ) (B- or N-edged) or armchair (BN pair-edged). Multi-layered h-BN structures were initially far more common prior to experimental optimization (Section 3.3.1) due to the ionic character of the B-N bond promoting interaction between neighboring sheets. This, in combination with the large energy penalties of homogenous bonds (B-B and N-N bonds), results in multiple layers often forming to stabilize the structure.[26]

1.4 Boron Nitride Nanotubes

As graphene sheets can be rolled into CNTs, so too can h-BN sheets be rolled into cylinders to form BNNTs, as shown in Fig. 1.3. As the B-N bond has an ionic character, an isolated single-walled boron nitride nanotube (SWBNNT) is difficult to obtain unlike a single-walled carbon nanotube (SWCNT). Prominent lip-lip interactions between adjacent h-BN layers often result in multi-walled boron nitride nanotubes (MWBNNTs). BNNTs possess chirality, referring to the axis about which the h-BN sheet is rolled relative to the h-BN lattice vectors. The chirality can be expressed in terms of a chiral vector (C_h), a linear combination of the h-BN lattice vectors (a_1 and a_2, Eq. 1.1), i.e., $na_1 + ma_2 = (n,m)$.

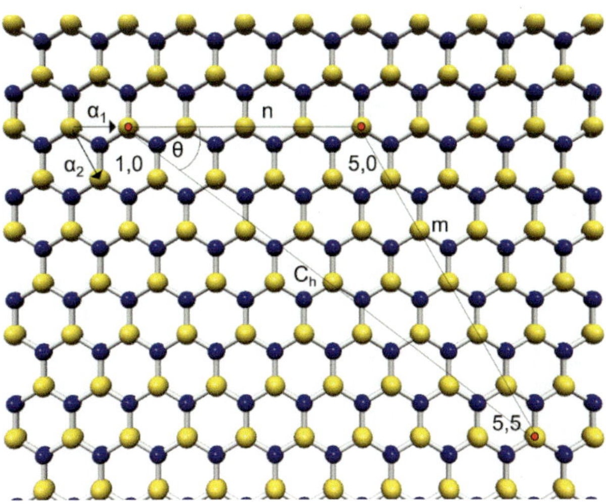

Figure 1.3 Construction of the chiral vector, Ch from lattice vectors, a_1 and a_2 to form a BNNT from h-BN. Atoms are colored yellow for boron and blue for nitrogen.

$$\{a_1, a_2\} = a\left\{\frac{\sqrt{3}}{2}, \pm\frac{1}{2}\right\}, \tag{1.1}$$

where a is the h-BN lattice constant (0.2504 nm). Subsequently, the values of n and m define the chirality of the BNNT and its chiral angle θ.

$$\sin(\theta) = \frac{m\sqrt{3}}{2\sqrt{n^2 + nm + m^2}} \qquad (0 \leq |\theta| \leq \frac{\pi}{6}) \tag{1.2}$$

For example, if the h-BN sheet in Fig. 1.3 is rolled along the direction α_1 such that (1,0) and (5,0) come into contact, the chiral angle (Eq. 1.2) will be zero, forming a ZZ (5,0) BNNT. In contrast, if the h-BN sheet was rolled such that (1,0) and (5,5) contact, θ will be 30°, forming an armchair (5,5) BNNT. Chiral angle θ values between 0° and 30° yield chiral BNNTs. An example of each configuration of BNNTs is presented in Fig. 1.4. While the chiral angle is extremely important for the resulting properties of SWCNTs, BNNTs are insulating materials irrespective of their chirality. Regardless of the chiral angle, the circumference of a BNNT is given by the magnitude of the chiral vector $|C_h| = a(n^2 + m^2 + nm)^{1/2}$ and hence the diameter $d = |C_h|/\pi$.

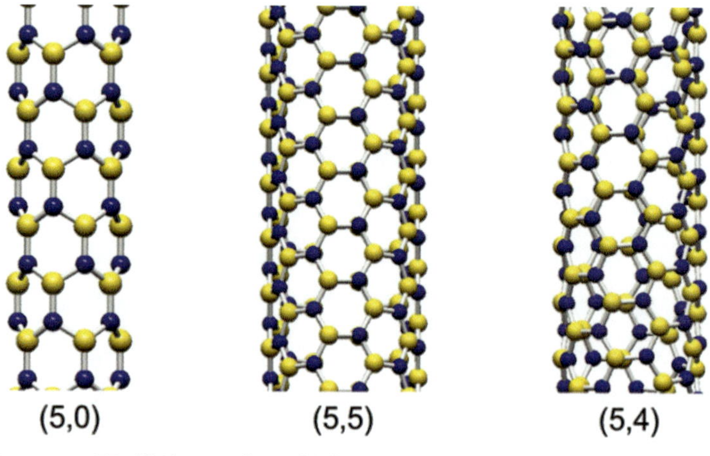

Figure 1.4 ZZ (5,0), armchair (5,5), and chiral (5,4) BNNTs. Atoms are colored yellow for boron and blue for nitrogen.

While h-BN nanosheets are terminated by an edge, BNNTs are terminated by a nanotube cap, initially theorized to be similar in structure to half of a BN fullerene. However, the structure of these caps can vary significantly for BNNTs, as discussed in Section 3.4.

Each phase of BN has unique properties and can be synthesized in a multitude of ways. As such, there has been a significant research effort to optimize and improve these synthesis routes, often targeting a specific resulting property for use in a tailored application. Each facet of the field of BN science and material engineering will be covered in this textbook in subsequent chapters. With this collection of information, it is the hope that BN materials can be further thrust to the forefront of materials science and the future generation of advanced materials produced at an industrial scale.

References

1. Balmain, W. H., XLVI., Observations on the formation of compounds of boron and silicon with nitrogen and certain metals. *The London, Edinburgh, and Dublin Philosophical Magazine and Journal of Science.* **1842,** *21* (138), 270–277.
2. Balmain, W. H., Bemerkungen über die Bildung von Verbindungen des Bors und Siliciums mit Stickstoff und gewissen Metallen. *J. Prakt. Chem.* **1842,** *27* (1), 422–430.
3. Dobrzhinetskaya, L. F.; Wirth, R.; Yang, J.; Green, H. W.; Hutcheon, I. D.; Weber, P. K.; Grew, E. S., Qingsongite, natural cubic boron nitride: The first boron mineral from the Earth's mantle. *Am. Mineral.* **2014,** *99* (4), 764.
4. Sōma, T.; Sawaoka, A.; Saito, S., Characterization of wurtzite type boron nitride synthesized by shock compression. *Mater. Res. Bull.* **1974,** *9* (6), 755–762.
5. Huang, J. Y.; Yasuda, H.; Mori, H., HRTEM and EELS studies on the amorphization of hexagonal boron nitride induced by Ball Milling. *J. Am. Ceram. Soc.* **2000,** *83* (2), 403–409.
6. Arenal, R.; Kociak, M.; Zaluzec, N. J., High-angular-resolution electron energy loss spectroscopy of hexagonal boron nitride. *Appl. Phys. Lett.* **2007,** *90* (20), 204105.

7. Sumiya, H.; Iseki, T.; Onodera, A., High pressure synthesis of cubic boron nitride from amorphous state. *Mater. Res. Bull.* **1983,** *18* (10), 1203–1207.

8. Gladkaya, I. S.; Kremkova, G. N.; Slesarev, V. N., Turbostratic boron nitride (t-BN) under high pressures and temperatures. *Journal of the Less Common Metals.* **1986,** *117* (1), 241–245.

9. Christensen, N. E.; Gorczyca, I., Optical and structural properties of III-V nitrides under pressure. *Phys. Rev. B.* **1994,** *50* (7), 4397–4415.

10. Xiao, J.; Du, J.; Wen, B.; Melnik, R.; Kawazoe, Y.; Zhang, X., Phase stability limit of c-BN under hydrostatic and non-hydrostatic pressure conditions. *J. Chem. Phys.* **2014,** *140* (16), 164704.

11. Pakdel, A.; Bando, Y.; Golberg, D., Nano boron nitride flatland. *Chem. Soc. Rev.* **2014,** *43* (3), 934–959.

12. Nagakubo, A.; Ogi, H.; Sumiya, H.; Kusakabe, K.; Hirao, M., Elastic constants of cubic and wurtzite boron nitrides. *Appl. Phys. Lett.* **2013,** *102* (24), 241909.

13. Kurdyumov, A. V.; Solozhenko, V. L.; Zelyavski, W. B., Lattice parameters of boron nitride polymorphous modifications as a function of their crystal-structure perfection. *J. Appl. Cryst.* **1995,** *28*, 540–545.

14. Lipp, A.; Schwetz, K. A.; Hunold, K., Hexagonal boron nitride: Fabrication, properties and applications. *J. Eur. Ceram. Soc.* **1989,** *5* (1), 3–9.

15. Le Godec, Y.; Martinez-Garcia, D.; Solozhenko, V. L.; Mezouar, M.; Syfosse, G.; Besson, J. M., Compression and thermal expansion of rhombohedral boron nitride at high pressures and temperatures. *J. Phys. Chem. Solids.* **2000,** *61* (12), 1935–1938.

16. Solozhenko, V. L., *J. Hard. Mater.* **1995,** *6*, 51.

17. Wentorf, R. H., Cubic form of boron nitride. *The Journal of Chemical Physics.* **1957,** *26* (4), 956–956.

18. Wentorf, R. H., Synthesis of the cubic form of boron nitride. *J. Chem. Phys.* **1961,** *34* (3), 809–812.

19. Bundy, F. P.; Wentorf, R. H., Direct transformation of hexagonal boron nitride to denser forms. *J. Chem. Phys.* **1963,** *38* (5), 1144–1149.

20. Vel, L.; Demazeau, G.; Etourneau, J., Cubic boron nitride: Synthesis, physicochemical properties and applications. *Mater. Sci. Eng. B.* **1991,** *10* (2), 149–164.

21. Wentorf, R. H., Condensed systems at high pressures. *J. Phys. Chem.* **1959,** *63* (11), 1934–1940.

22. Tani, E.; Sōma, T.; Sawaoka, A.; Saito, S., A stability of wurtzite type boron nitride under high pressure. *Jpn. J. Appl. Phys.* **1975,** *14* (10), 1605–1606.

23. Solozhenko, V. L., On phase diagram of boron nitride. *Doklady Akademii Nauk SSSR.* **1988,** *301* (1), 147–149.

24. Will, G.; Nover, G.; von der Gönna, J., New experimental results on the phase diagram of boron nitride. *J. Solid State Chem.* **2000,** *154* (1), 280–285.

25. Solozhenko, V. L.; Turkevich, V. Z.; Holzapfel, W. B., Refined phase diagram of boron nitride. *J. Phys. Chem. B.* **1999,** *103* (15), 2903–2905.

26. Blase, X.; De Vita, A.; Charlier, J. C.; Car, R., Frustration effects and microscopic growth mechanisms for BN nanotubes. *Phys. Rev. Lett.* **1998,** *80* (8), 1666–1669.

Chapter 2

Properties of BN Nanomaterials

2.1 Introduction

Material properties drive our interest and research effort, irrespective of the material in question. For instance, once c-BN was identified as a material with a hardness rivaling that of a diamond but at a lower cost. Researchers then focused on synthesizing quantities of c-BN that would enable its commercial uptake in applications such as abrasives and coatings. Similarly, the chemical stability and insulating properties of h-BN are very attractive for hybrid applications with conducting materials such as graphene. This possibility drove efforts in synthesizing defect-free, large-area h-BN films. For similar reasons, significant research efforts have been directed toward reliable and scalable BNNT synthesis methods. In this chapter, we examine key mechanical, thermal, optical, and electronic properties for each phase of BN. We highlight key studies that demonstrate how to manipulate or enhance these properties and provide insight into how the understanding of these materials has evolved over time using a wide array of techniques.

2.2 Mechanical Properties

The hardness of a material is a measure of its ability to scratch or indent a surface. Multiple hardness values are reported in

Boron Nitride Nanomaterials: Properties, Fabrication, and Applications
Ben McLean and Alister J. Page

ISBN 978-981-4968-23-2 (Hardcover), 978-1-003-31448-6 (eBook)
www.jennystanford.com

Table 2.1, along with key mechanical properties for each BN phase. Knoop hardness is a method involving a pyramidal diamond indenting tip forced into a surface for a predetermined period of time with a set load. The length of the indent in the surface is measured and this with the load allows the Knoop hardness to be calculated. The Vickers hardness test also typically uses a pyramidal diamond indenter, though the methods of calculation are independent of the size of the indenter, requiring only the force applied to the diamond upon indentation and the surface area of the indentation. Nanohardness, reported for c-BN in Table 2.1, is measured *via* a similar method but with a much smaller indentation diameter, typically about 150 nm. This method was developed in the 1970s to measure a smaller volume of synthetic material for use in modern applications.[1–3] Many studies report different hardness values. During indentation, there is elastic and a plastic component for the deformation of a material. For plastic deformation, the displacement of the material remains relatively constant after the removal of the indenter. For elastic deformation, the material exhibits a certain degree of recovery, reducing the observed displacement after indentation, compared to when the load is applied. Nanoindentation measurements help alleviate errors in the measurement due to elastic recovery. Elastic recovery can also be mistaken for film bending. Further, the surface must be sufficiently homogenous, and also the measurement is taken as far from the edge as possible to minimize error.

Young's modulus quantifies the relationship between tensile stress and axial strain in the material during elastic deformation. Graphically, it is the slope of the linear elastic region of a typical stress-strain curve. The shear and bending moduli are related quantities, defined as the ratio of stress to strain when the material is subjected to shear stress and torsional stress, respectively. Poisson's ratio is the ratio of transverse strain to axial strain when the material is subjected to stress along a single axis. The bulk modulus is a measure of a material's resistance to a change in volume with applied pressure. While this is used more commonly in characterizing fluids, it can be measured by applying pressure during powder diffraction measurements for solid-phase materials. Fracture toughness represents the resistance of a material under stress to fracture once a flaw or a crack has formed. This is proportional to the energy consumed during plastic deformation. The fracture toughness, as with most moduli presented in Table 2.1, is significantly influenced by the presence of defects and grain boundaries in a material.

2.2.1 Mechanical Properties of c-BN

Being structurally analogous to diamond, c-BN shares diamond-like properties, as characterized by a wide range of techniques. Following the discovery of c-BN,[4] it was claimed that c-BN was hard enough to scratch a diamond. Naturally, this led to the characterization of its mechanical properties and its use in hard ceramic tools (more in Chapter 4), though at first, many investigations of c-BN properties were reported *via* patents.[5–12] Once the use of c-BN as a replacement for diamond became more commonplace, efforts to reduce synthesis costs and simultaneously increase the hardness of the synthesized material multiplied. A key observation in this respect was that the substrate can also influence the measurement of mechanical properties. Mirkarimi *et al.*[13] characterized c-BN films deposited by ion-assisted sputtering (Section 3.2.2) on Si and showed that the thickness of the c-BN results in the load-displacement plot deviating from linearity, in agreement with a model proposed by Hainsworth and Page indicating that the Si substrate deforms and plays a role in the properties of the grown c-BN.[14,15] As a result, the development of stress in the c-BN film is common,[16] and significant research effort has been made to optimize the growth process to reduce film stress and increase the content of c-BN compared to h-BN in the film produced. New c-BN polymorphs are investigated theoretically to uncover the mechanically superior form of this hard material.[17,18] Sumiya *et al.*[19] reported that the hardness and fracture strength depended strongly on the c-BN grain size and the c-BN content of the film. They found that fracture strength increased with temperature and this attributed to plastic deformations that relaxed the stress during load application. The balance of synthesis conditions was important to minimize the amount of compressed h-BN and the coarseness of the c-BN grains. With finer grains, the material hardness increases as per the Hall–Petch effect.[20,21] Dub *et al.*[22] also saw a fine-grained (100 nm) polycrystalline c-BN product containing clear grain boundaries (produced by conversion of h-BN under extreme conditions) that exhibit a high Young's modulus and fracture strength (Fig. 2.1a). As polycrystalline c-BN started to exhibit enhanced mechanical properties to single-crystal c-BN, research efforts began to switch to synthesizing highly nanostructured c-BN. For instance, Dubrovinskaia *et al.*[23] reported the synthesis of an aggregated c-BN nanocomposite that enhanced hardness by 100% compared to single-crystal c-BN (Fig. 2.1c). The ultrafine grains (14 nm) allowed the measured Vickers hardness to be 85 GPa due to the Hall–Petch effect. The hardness was observed to drop below 14 nm grain size in accordance with an inverse Hall–Petch effect, attributed to grain boundary sliding. This value for c-BN was said to be the limit for the Hall–Petch effect, observed for other metals and alloys.[24–27] Solozhenko *et al.*[28] transformed h-BN with an "ideal random layer" into an ultra-hard nanostructured c-BN that also exhibited a Vickers hardness of 85 GPa for very fine grain size (Fig. 2.1d). Tian *et al.*[29] also produced an ultra-hard c-BN and observed fine twin domains of an average thickness of approximately 3.8 nm (Fig. 2.1b). The Vickers hardness of this material exceeds that of synthetic diamond (100 GPa) and has a fracture toughness greater than 12 MPa $m^{1/2}$.

2.2.2 Mechanical Properties of h-BN

h-BN exhibits a very high Young's modulus around 1,000 GPa.[31] Monolayer h-BN was predicted to also possess a high Young's modulus as a monolayer between 716 and 977 GPa.[32–37] However, initial experiments measuring chemical vapor deposition (CVD)-synthesized h-BN showed it to be far lower than these predictions (200–500 GPa, Fig. 2.2b).[38] Later Kim *et al.* synthesized large-area multilayer h-BN on Fe that exhibited an extremely high Young's modulus, i.e., 1,160 GPa.[39] Interestingly, unlike graphene, h-BN mechanical properties do not depend on having a limited number of layers in the structure, according to experiment. For instance, Falin *et al.*[40]

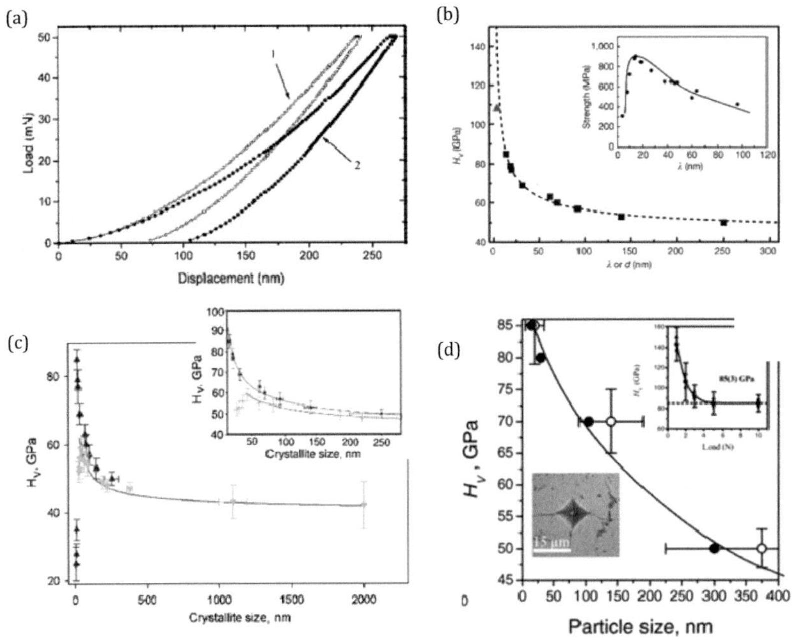

Figure 2.1 (a) Load displacement curves produced by Dub *et al.* show the enhanced response of fine-grain c-BN compared to coarse grain.[22] Reprinted with permission from Dub, S. N. *et al.* Mechanical properties of polycrystalline cBN obtained from pyrolytic gBN by direct transformation technique. *High Press. Res.* 2006, 26 (2), 71–77. Copyright 2006 Taylor & Francis Ltd. http://www.tandfonline.com. (b) Vickers hardness as a function of twin thickness for Tian *et al.*'s nanotwinned c-BN with inset showing optimum twin thickness for maximum strength.[29] Reprinted by permission from Springer Nature Customer Service Center GmbH: Springer Nature Tian, Y. *et al.* Ultra-hard nanotwinned c-BN. Nature 2013, 493 (7432), 385–388. Copyright 2013. (c) Vickers hardness as a function of crystallite size measured by Dubrovinskaia *et al.* for c-BN (blue) and aggregated BN nanocomposites (red).[23] Reprinted from Dubrovinskaia, N. *et al.* Super-hard nanocomposite of dense polymorphs of BN: Non-carbon material has reached diamond hardness. *Appl. Phys. Lett.* 2007, 90 (10), 101912. Copyright 2007 AIP Publishing. (d) Vickers hardness as a function of c-BN grain size measured by Solozhenko *et al.* using X-ray diffraction (XRD) (closed circles) and transmission electron microscope (TEM) (open circles).[30] Reprinted from Solozhenko, V. L. *et al.* Mechanical properties of ultra-hard nanocrystalline cubic boron nitride. *J. Appl. Phys.* 2019, 127 (7), 075107. Copyright 2019 AIP Publishing.

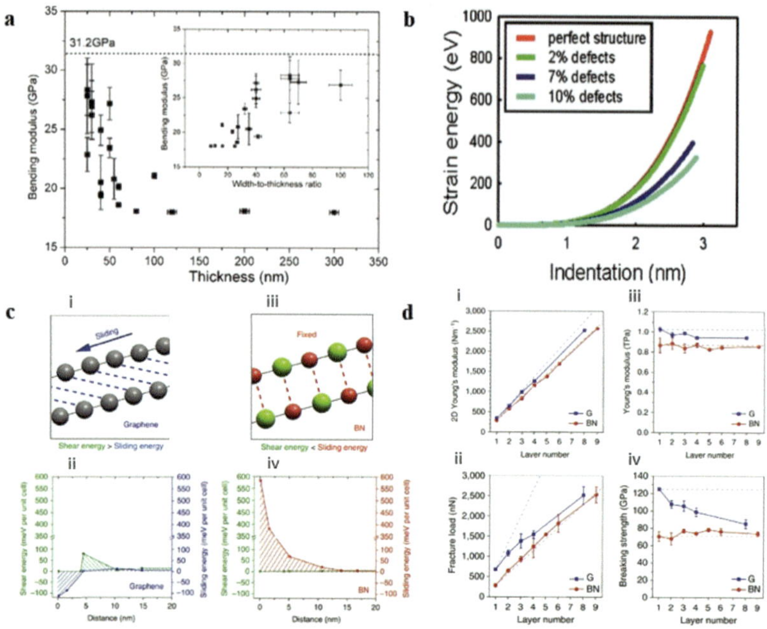

Figure 2.2 (a) Bending modulus of h-BN as a function of the sheet thickness and (inset) width to thickness ratio.[41] Li, C. *et al.* Thickness-dependent bending modulus of h-BN nanosheets. Nanotechnology 2009, 20 (38), 385707 © IOP Publishing. Reproduced with permission. All rights reserved. (b) Strain energy of h-BN sheet measured by Song *et al.* for different defect concentrations.[38] Reprinted with permission from Song, L. *et al.* Large scale growth and characterization of atomic hexagonal boron nitride layers. *Nano Lett.* 2010, 10(8), 3209–3215. Copyright 2010 American Chemical Society. (c) (i, ii) Diagrams representing sliding behavior as a result of plots of shear strain energy between (iii) graphene and (iv) h-BN sandwich structures showing no negative shear energy for h-BN.[40] Distributed under a Creative Commons CC BY license. (d) (i) 2D Young's modulus, (ii) volumetric Young's modulus, (iii) fracture load, and (iv) breaking strength comparisons by Falin *et al.* between graphene and h-BN as a function of layer number.[40] Distributed under a Creative Commons CC BY license.

measured Young's modulus of 865 GPa and fracture strength of 70.5 GPa showing little change between 1 and 9 layers (Fig. 2.2c,d). This is markedly different behavior compared to that established for mono- and few-layer graphenes, and has been attributed to the difference in sliding behavior between layers for h-BN and graphene. Mirnezhad *et al.*[36] in contrast used density functional theory (DFT) calculations to show that multilayer h-BN is weaker than monolayer h-BN, depending on the stacking order. Li *et al.*[41] did show that the number of layers also influences the bending modulus, however *via* stacking faults following exfoliation (Fig. 2.2a).

2.2.3 Mechanical Properties of BNNTs

Compared to the planar phases of BN, measuring the mechanical properties of BNNTs can be problematic and this has led to a wide range of reported Young's moduli, as is evident in Fig. 2.3.

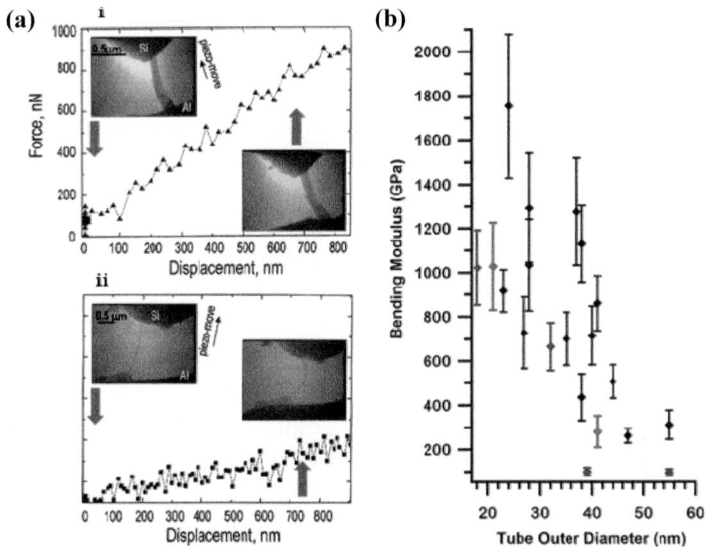

Figure 2.3 (a) Force vs. displacement measured using a combined atomic force microscopy (AFM)-TEM setup by Golberg *et al.* for thick (i) and narrow (ii) BNNTs with insets showing initial and bent configurations of the tubes.[44] Reprinted with permission from Golberg, D. *et al.* Direct force measurements and kinking under elastic deformation of individual multiwalled boron nitride nanotubes. *Nano Lett.* 2007, 7 (7), 2146–2151. Copyright 2007 American Chemical Society. (b) Bending modulus calculated by Tanur *et al.* as a function of outer tube diameter using a simply supported beam model (black), double clamped beam model (red), and mixed support beam model (blue).[47] Reprinted with permission from Tanur, A. E. *et al.* Diameter-dependent bending modulus of individual multiwall boron nitride nanotubes. *J. Phys. Chem. B* 2013, 117 (16), 4618–4625. Copyright 2013 American Chemical Society.

For instance, Chopra *et al.*[42] measured Young's modulus inside a TEM *via* the resonant frequency of a cantilevered BNNT by driving it with a forced oscillator and measuring the thermal vibration amplitude. Young's modulus here was measured to be 1220 GPa for MWBNNTs. Suryavanshi *et al.*[43] found that MWBNNTs with diameters 34–94 nm exhibited Young's modulus of 722 GPa on average, using a TEM-based method in which a cantilevered tube was made to resonate *via* an applied electric field. Golberg *et al.*[44] used an integrated TEM-AFM piezo-driven holder to accurately record force, piezo displacement curves (Fig. 2.3a). By compressing the BNNTs between a piezo-moveable Al wire and a Si cantilever, they measured Young's modulus to be 500–600 GPa. Similarly, a modified TEM-AFM setup employed by Arenal *et al.*[45] allowed for Young's moduli of SWBNNTs to be measured. The determination of Young's modulus requires the wall thickness to be known and depending on how that is defined, Young's modulus varies significantly. If the wall thickness is assumed to be the same as the interlayer spacing of graphite, 0.34 nm; Young's modulus is 250 GPa, if the thickness is 0.07 nm as per some analytical approaches,[46] Young's modulus is 1110 GPa. A separate study of MWBNNT Young's moduli reported a similarly wide range *via* atomic force microscopy (AFM) measurements, 1,800 ± 300 GPa (Fig. 2.3b).[47] Theoretical works utilizing classic potentials[48] and tight binding[49] place Young's moduli of SWBNNTs between 800 and 1000 GPa. Table 2.1 demonstrates Young's modulus of BNNT that is, in reality, uncertain with varying values reported due to the range of experimental techniques and the difference in BNNT quality between studies.

Table 2.1 Mechanical properties of BN phases

Mechanical property	c-BN	h-BN	BNNT
Knoop Hardness (GPa)	~45[50–53] 60–70[54] 58–76[55] >100[29]		
Vickers Hardness (GPa)	49–54[56] 62[52] 60–75[57] 75[58]		

Mechanical property	c-BN	h-BN	BNNT
Young's Modulus (GPa)	800–900[59] 909[60]	279.2[35] 750[37] 785[61] 811[33] 829[36] 865[40] 881[32]	225[62] 284–310[49] 350[63] 500–600[44] 450–680[64] 780[65] 844–1006[66] 505–1031[43] 1070[67] 250–1110[45] 982–1137[48] 1220[42] 25–1343[68] 1800[47]
Poisson Ratio		0.210[61] 0.213[36] 0.2176[35] 0.297[37]	0.13–0.16[48] 0.211[69] 0.56–0.57[65] 1.45–1.53[66]
Shear Modulus (GPa)	407[30]	3[70] 3–4[71] 6.61[40] 7.7[33] 3–9[72] 7–9[73]	0.507–5.928[48] 1.4[63] 7[47]
Bending Modulus (GPa)		31.2[41]	760[47]
Fracture Toughness ($MPa\ m^{1/2}$)	2.8[59] 3.0[52], 3.5–5[59] 5.0[56],	70.5 GPa[40] 133.3 GPa[32]	8 GPa[62] 9.1–15.5 GPa[74] 30 GPa[68]

2.3 Thermal Properties

Current-generation silicon-based electronic devices employ SiO_2 dielectrics. The thermal conductivity of SiO_2 around room temperature is 1 W m^{-1} K^{-1}.[75] All forms of BN, particularly h-BN far exceed this thermal conductivity (Table 2.2) and thus offer great potential for use as a substrate in next-generation thin-film applications. Thermal conductivity is simply the ability of a material to conduct heat. The value represents the quantity of thermal energy that flows through a material. There are several ways to measure thermal conductivity. Earlier methods involved heating the substrate and measuring the response of the material with a thermometer. Pulse methods were used with pulses shorter than the time for a discernible temperature change to be measured, thus reducing the energy lost to the surrounding medium or through the boundary conditions.[76] Thermo-reflectance techniques use ultra-fast laser pulses to limit thermal penetration, hence can make measurements on nanoscale thin films.[77] However, volumetric heat capacity is required, thus making it difficult for novel 2D materials. For graphene, Raman spectroscopy is a successful method of measuring the thermal conductivity as the Raman G peak frequency depends on the excitation laser power and the measurable G peak temperature coefficient.[78–82] Due to the comparably weaker intensities of the Raman peaks in a few layers of h-BN, this method is not as easily adopted. Some micro-Raman spectroscopic methods have seen some success as the E_{2g} peak frequencies vary linearly with temperature.[83] Other methods have involved using microbridge devices that electrically heat a U-shaped metal line and measure the electrical resistance of adjacent metal lines to determine the resulting temperature rise. This method has been used for graphene[84] and is used to calculate the thermal resistance of the h-BN in the device.[85] For BNNTs, Chang *et al.*[86] placed individual tubes in a microscale test fixture using a piezo-driven manipulator inside a scanning electron microscope (SEM). BNNTs were suspended between independent heat sinks. As the power was supplied to a heater, the resistance changes in both the heater and integrated Pt film resistors allowed the changes in temperature to be calculated and subsequently the thermal conductivity.

Table 2.2 Thermal properties of BN phases

Thermal property	c-BN	h-BN	BNNT
Thermal Conductivity ($W\ m^{-1}\ K^{-1}$)	880[92]	100[102] rt (10–20 nm) 227–280[83] rt (few layer) 250[85] rt (5 layer) 360[85] rt (11 layer) 390[76] rt (bulk) 585[103] rt (monolayer) 600[94] rt (monolayer) 751[95] rt (monolayer)	175[86] (natural) 300[86] (isotopic enhancement)

2.3.1 Thermal Properties of c-BN

Since the measurement of diamond's thermal conductivity, it has been classified as the most thermally conductive bulk material at room temperature.[87] Additionally, it was shown that a synthetic isotopically pure (C^{12}) diamond enhances the thermal conductivity due to ^{13}C scattering phonons more strongly and decreases the measured thermal conductivity.[88,89] Novikov *et al.* showed that this also holds for c-BN, with isotopic purity of ^{11}B or ^{10}B increasing the theoretical thermal conductivity significantly, up to 125 times.[90,91] Recently, Chen *et al.*[92] prepared four sets of c-BN crystals containing natural N and different B isotope compositions as confirmed by Raman and time-of-flight second ion mass spectrometry (TOF-SIMS). A laser pump-probe technique was combined with spatial and temporal thermoreflectance to measure thermal transport. Natural c-BN (78.3% ^{11}B) exhibited a thermal conductivity of 880 W m^{-1} K^{-1}. Incredibly, when ^{11}B was enriched to 99.2%, the thermal conductivity increased to 1,660 W m^{-1} K^{-1} and 1,650 W m^{-1} K^{-1} for 99.3% ^{10}B, due to the removal of phonon-isotope scattering that occurs with natural c-BN. Figure 2.4 compares the thermal conductivity of natural c-BN and isotopically enhanced c-BN to a range of semiconductors and boron-containing materials, highlighting the different forms of phonon scattering processes.

2.3.2 Thermal Properties of h-BN

The phenomenon of isotopic enhancement is also present for h-BN. Lindsay *et al.*[94] calculate room temperature thermal conductivity of more than 600 W m^{-1} K^{-1} for monolayer h-BN using the phonon Boltzmann transport equation (Fig. 2.5), agreeing well with the experiment.[76] They show that thermal conductivity is enhanced significantly by the isotopic purity of the sample. In addition, the thermal conductivity of h-BN increases markedly for an increase in crystallite size with isotopic enhancement. Increasing the crystallite size to 1 μm sees a 40% increase while 10 μm sees a 200% increase. Cai *et al.*[95] reported a room temperature thermal conductivity of 751 W m^{-1} K^{-1} for monolayer h-BN as measured *via* Raman spectroscopy. In this study, the thermal conductivity was also reduced with an increasing number of layers of h-BN, decreasing to 646 W m^{-1} K^{-1} for two layers and 602 W m^{-1} K^{-1} for three layers. This trend agreed with classical molecular dynamics (MD) simulations performed in the same paper, attributed to phonon branching due to the interlayer interactions. For multilayer h-BN, Jo *et al.*[85] showed that the thermal conductivity starts to approach that of bulk h-BN and that the presence of a polymer layer on the surface decreased few-layer h-BN thermal conductivity, notable for future application.

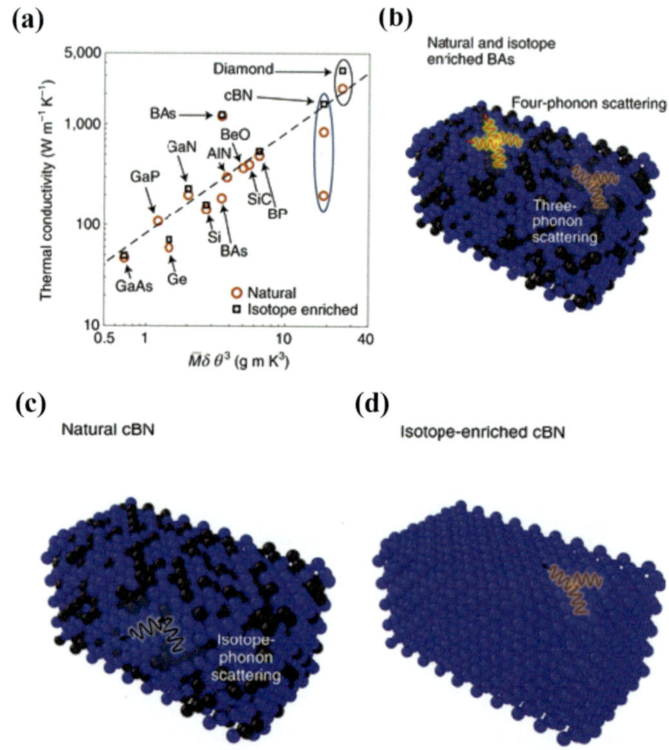

Figure 2.4 (a) Room-temperature thermal conductivity values as a function of Slack scaling parameter of common semiconductors as well as c-BN and diamond. (b–d) Schematics describe the different phonon scattering mechanisms in crystals with natural and enhanced isotopic content. Negligible influences are observed for isotopically enhanced c-BN from four-phonon processes. (see ref. 93 and references therein).[93] Reprinted by permission from Springer Nature Customer Service Center GmbH: Springer Nature Giri, A. *et al.* Achieving a better heat conductor. *Nat. Mater.* 2020, 19 (5), 482–484. Copyright 2020.

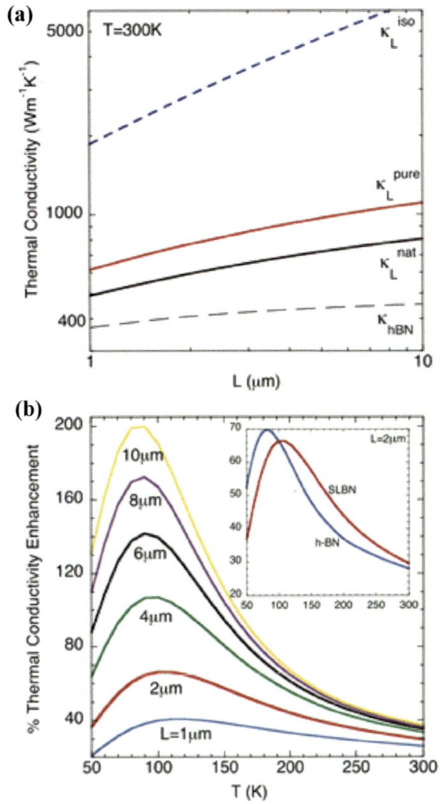

Figure 2.5 (a) Thermal conductivity as a function of L, a measure of the length between crystallite boundaries in the transport direction as calculated by Lindsay *et al.* for different regimes of phonon scattering in monolayer h-BN; boundary and isotopic (blue), boundary and phonon-phonon (red), all three (black), and that for bulk h-BN (black, dashed).[94] (b) Percentages of thermal conductivity enhancement as a function of temperature and L for isotopically pure monolayer h-BN.[94] Reprinted with permission from Lindsay, L. *et al.* Enhanced thermal conductivity and isotope effect in single-layer hexagonal boron nitride. *Phys. Rev. B* 2011, 84 (15), 155421. Copyright 2011 by the American Physical Society.

2.3.3 Thermal Properties of BNNTs

At room temperature, MWBNNTs have a thermal conductivity of approximately 175 W m^{-1} K^{-1}, somewhat smaller than that of a multi-walled carbon nanotube (MWCNT) (250 W m^{-1} K^{-1}).[86]

Chang *et al.*[86] showed however that with isotopic enhancement (99.5% ^{11}B), this thermal conductivity increases by 50% to rival that of an MWCNT (300 W m^{-1} K^{-1}), though it is worth noting that this is an expensive process. Additionally, with the increasing outer diameter, the thermal conductivity of the BNNTs decreased, attributed to increased phonon scattering between walls. The Green's function[96] approach reported by Stewart *et al.*[97] investigated the source of Chang *et al.*'s observations, however, considered an SWBNNT over MWBNNTs. Their predicted data agreed well with Chang *et al.*'s measurements above 150 K, indicating that in MWBNNTs, heat conduction may occur primarily in the outer nanotube shell, something proposed by a prior study for MWCNTs.[98] Using a classical Tersoff-derived potential, Jiang *et al.*[99] investigated phonon scattering in SWBNNTs in more detail and found the thermal conductivity to be influenced by the screw symmetry of the SWBNNT.

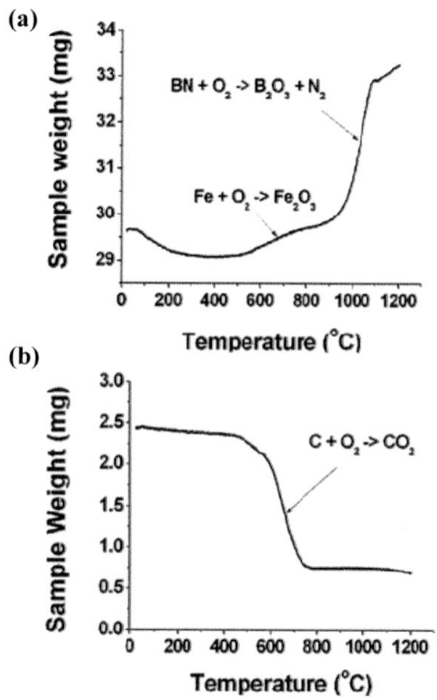

Figure 2.6 Sample weight changes in (a) BNNT and (b) CNT samples as a function of temperature during thermogravimetric analysis (TGA) in air measured by Chen *et al.* showing BNNTs superior resistance to oxidation.[101] Reprinted from Chen, Y. *et al.* Boron nitride nanotubes: Pronounced resistance to oxidation. *Appl. Phys. Lett.* 2004, 84 (13), 2430–2432, with the permission of AIP Publishing.

The thermal properties of vertically aligned BNNT films were studied by Belkerk *et al.*[100] using a photothermal technique that employed a Pt thin-film thermometer sputter-deposited on the SiO_2 substrate prior to growth; the temperature of the Pt thermometer changed with laser pulsing and allowed for the thermal conductivity of the BNNTs to be measured during the growth process. As the thickness of the MWBNNTs increased from 10 μm to 28 μm, the thermal conductivity increased from 55 W m^{-1} K^{-1} to 170 W m^{-1} K^{-1}. Apparent medium theory translated these thermal conductivities to 470 W m^{-1} K^{-1} and 2,400 W m^{-1} K^{-1} for the solid parts (BNNTs) of the 10 μm and 28 μm-thick BNNT films, respectively. Largely, BNNTs exhibit high thermal conductivity albeit lower than CNTs. However, importantly for application, BNNTs show pronounced resistance to oxidation at higher temperatures in comparison to CNTs. Chen *et al.*[101] showed that BNNTs oxidized at 800 °C while CNTs did at 400 °C (Fig. 2.6).

2.4 Optical Properties

Optical characterization using methods such as spectroscopy and microscopy provides key information regarding the structure of a material, such as the internal bonding structure or the morphology of the material itself. Typically, these will have an influence on the properties of the material and hence their utility in target applications. Spectroscopic methods have been used to analyze the structure and composition of BN materials since the first synthesis of c-BN by Wentorf,[4] whereby spectrographic analysis detected impurities in the product. Today, Raman spectroscopy can be used *in situ* and *ex situ* to confirm the identity and purity of a BN product,[104] and can also be used to monitor atomic changes in the material in response to a stimulus such as changes in pressure and temperature.[105] Infrared (IR) and Raman spectroscopy are typically used in tandem with each other and with other characterization techniques, predominantly photoluminescence (PL) spectroscopy (whereby lasers induce absorption and probe the subsequent emission of photons from a material). IR, for example, is not effective for distinguishing graphite and CNTs due to poor IR transmittance and the proximity of the active modes. BN materials, on the other hand, having polar/ionic B-N bonds, have enhanced IR absorption. Raman scattering intensities are lower than they are for CNTs, due to their large band gap (see Section 2.5), which makes Raman scattering non-resonant. Interpretation of either spectrum is possible, however, with accurate knowledge of the phonon frequencies and how these may be influenced by the structure of the material, i.e., layers of h-BN or diameter/chirality of BNNTs.[106] Table 2.3 presents the optical properties refractive index, boundary modes, optical band gap and Raman excitations for each BN phase.

Table 2.3 Optical properties of BN phases

Optical Property	c-BN	h-BN	BNNT
Refractive Index	2.117[53]	1.96[131] 2.17[135] 2.22[134]	
Brillouin Zone Boundary Modes (cm^{-1})	LO 1232 TO 1000[53] LO 1269 TO 1027[110] LO 1304 TO 1056[107] LO 1306.4 TO 1055.6[113]	LO 823 TO 1382[149] LO 823.5 TO 1365[135] LO 828 TO 1367[148]	LO 805 TO 1380[156] LO 809 TO 1369[152] LO 810 TO 1380[155] LO 813 TO 1389[157]
Optical Band Gap (eV)	6.36 (Evans)	6.25 (bulk) 7.3[144] (monolayer)	5.75[163] 5.8[179]
Raman excitations mode (cm^{-1})		E_{2g} 1366[150] (bulk) 1367[148] 1367[182] (monolayer) 1370[183] (monolayer) 1368–1370[150] (monolayer) 1382[149]	A_1 1356[156] 1367[152] 1370[158]

2.4.1 Optical Properties of c-BN

Gielisse *et al.*[53] provided the first report of the optical properties of c-BN such as the refractive index (2.117), absorption, optical frequencies, and more, *via* IR spectroscopy. Results agreed with the following work by Brafman *et al.*[107] who employed Raman spectroscopy. Subsequent peak assignments reported by Chrenko[108] (FTIR spectroscopy) were not in agreement with these earlier reports, but were in agreement with observations of Eremets *et al.*[109] More recent theoretical work by Cai *et al.*,[110] using linear response theory, supported Gielisse *et al.*'s longitudinal optical (LO) and transverse optical (TO) peak frequencies.[110] The disagreement here is illustrated in Fig. 2.7, which shows the IR spectrum for a c-BN film synthesized by Ullmann *et al.*[111] The main feature of the IR spectrum here is the TO phonon peak at 1,087 cm^{-1}. With decreasing c-BN content (predominantly sp^3-bond character), the h-BN peaks at 800 cm^{-1} and 1,370 cm^{-1} (due to increasing sp^2 bond character) become more pronounced and the c-BN peaks become dampened.[112] High-quality c-BN produced by Zhang *et al.*[113] and analyzed by Raman spectroscopy showed two clear peaks for c-BN at 1,055.6 cm^{-1} (TO) and 1,306.4 cm^{-1} (LO), with no characteristic h-BN peaks.

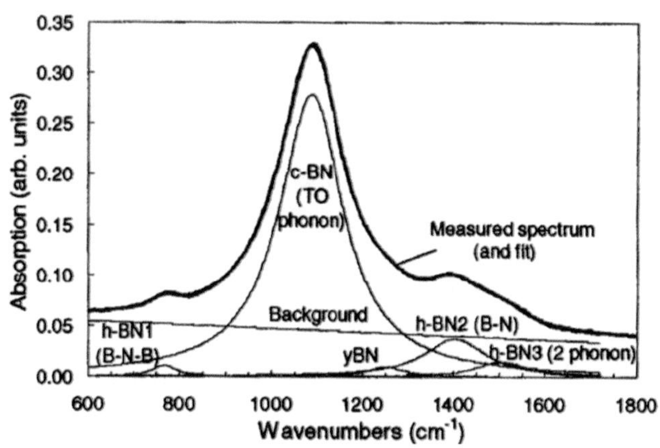

Figure 2.7 Deconvolution of the measured IR spectrum of a BN film with high c-BN content using five Lorentzian fitting curves.[111] Reprinted from Ullmann, J. *et al.* Effects of MeV ion irradiation of thin cubic boron nitride films. *J. Appl. Phys.* 1998, 83 (6), 2980–2987, with the permission of AIP Publishing.

Using either IR or Raman spectroscopy, it is possible to estimate the c-BN content of a film. For example, the relative height of the IR peaks at approximately 1,080–1,400 cm^{-1} gives the fraction of c-BN in the film, though these peaks can be shifted by stress[114] or temperature, film thickness, and stoichiometry.[115] A similar strategy is commonly reported for reporting the structural quality of mono- and few-layer graphene structures (i.e., the $I_D:I_G$ ratio); this, however, relies on Raman spectroscopy. Raman peaks of c-BN can broaden with defects or as the crystallite size decreases[116–118] as this influences the phonon dispersion in the material and hence the Raman scattering intensity. Figure 2.8a demonstrates in particular how the c-BN or h-BN film thickness influences the IR peak positions. Yokoyama *et al.*[119] also showed that the ratio of absorption coefficients can be used to obtain the ratio of c-BN:h-BN. In an extension of this approach, Jäger *et al.*[120] proposed a method using IR absorption measurements to calculate the volume fractions of hexagonal and cubic phases in a BN film, accounting for their differences in density and different IR band strengths (Fig. 2.8b).

When considering practical applications, the most important optical property of a material is arguably the band gap. Early theoretical works predicted c-BN to have an indirect, wide band gap.[121–123] Various methods have been applied to estimate the band gap of c-BN such as UV-photoemission measurements,[124] UV absorption spectra,[108] and soft X-ray spectra,[125] giving values between 5.8 eV and 6.4 eV. Evans *et al.*[122] employed synchrotron-based luminescence excitation spectroscopy to produce a value of approximately 6.36 eV (Fig. 2.9). This agrees with G_0W_0 calculations by Jiang and Blaha[126] (6.36 eV) though disagrees slightly with similar GW calculations by Gao[127] (6.18 eV) and Klimeš *et al.*[128] (6.66 eV). A comprehensive investigation into the optical properties of c-BN was carried out by Tararan *et al.*[129] employing EELs, luminescence spectroscopy, interferometry, and many-body perturbation theory. Interestingly, they argue that a band gap obtained from optical measurements should be compared to the smallest direct gap, rather than the indirect gap, in the band structure. For the previous GWA calculations by Klimeš *et al.*,[128] this amounts to 11.29–11.66 eV. Tararan *et al.*[129] propose a c-BN band gap of just over 10 eV, attributing prior measurements around 6 eV to be misinterpretations of optically active defects in measured c-BN samples.

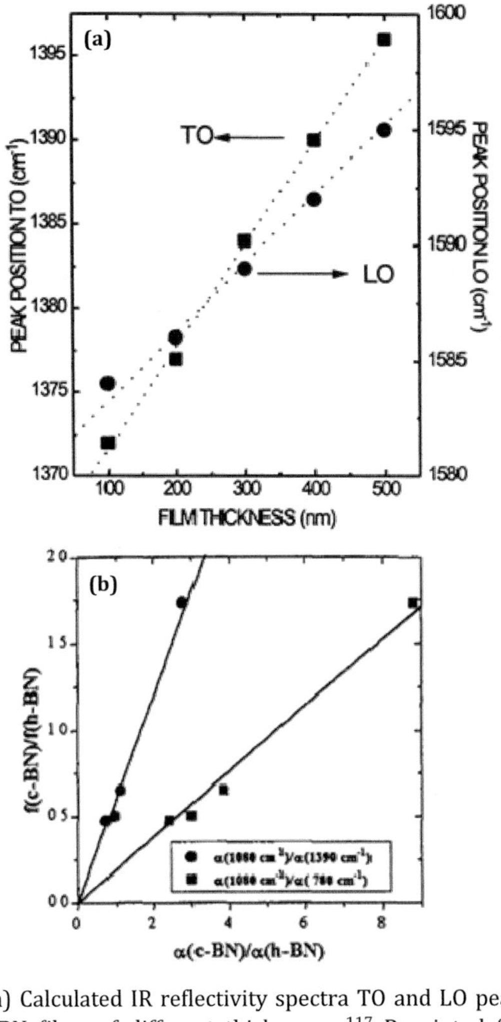

Figure 2.8 (a) Calculated IR reflectivity spectra TO and LO peak positions for c-BN/h-BN films of different thicknesses.[117] Reprinted from Ben el Mekki, M. *et al.* Structure investigation of BN films grown by ion-beam-assisted deposition by means of polarised IR and Raman spectroscopy. *Surf. Coat. Technol.* 1999, 116–119, 93–99. Copyright 1999, with permission from Elsevier. (b) Correlation between the adsorption coefficient ratios (α(1,080 cm^{-1})/α(h-BN)) and calculated real volume fraction ratio of c-BN:h-BN for the 1,390 cm^{-1} and 780 cm^{-1} absorption modes.[120] Reprinted from Jäger, S. *et al.* Infrared spectroscopic investigations on h-BN and mixed h/c-BN thin films. *Thin Solid Films* 1994, 245 (1), 50–54. Copyright 1994, with permission from Elsevier.

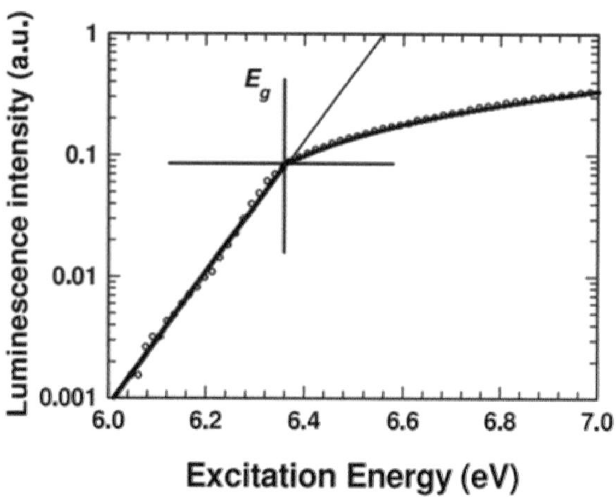

Figure 2.9 The band-edge luminescence excitation of c-BN as measured by Evans *et al.* using luminescence excitation spectroscopy showing the band gap, 6.36 eV.[122] Evans, D. A. *et al.* Determination of the optical band gap energy of cubic and hexagonal boron nitride using luminescence excitation spectroscopy. *J. Phys. Condens. Matter* 2008, 20 (7), 075233 © IOP Publishing. Reproduced with permission. All rights reserved.

2.4.2 Optical Properties of h-BN

The refractive index of h-BN has been reported over a range of 1.5–2.2.[109, 130–134] While this is comparable to the value reported for c-BN (see Section 2.3.1), there is a disagreement between techniques such as ellipsometry, theory, and transmittance, summarized by Lee *et al.*[135] Additionally, McKenzie *et al.*[130] showed the refractive index of h-BN to vary with film thickness. There has been similar confusion regarding the optical band gap of h-BN; early optical measurements of h-BN reported an optical band gap much lower than the bulk value due to the thickness of the films (> 6 nm).[136, 137] For instance, h-BN thin films studied by Zunger *et al.*[138] exhibited a direct band gap of 6.2 eV. In contrast, many following works predicted an indirect band gap[123, 139, 140] but important luminescence observations of the deep-UV absorption of a high-quality h-BN crystal by Watanabe *et al.*[141] definitively proved h-BN to be a direct band gap material. Due to these exciton absorption bands, the band gap was estimated to be

5.971 eV, in agreement with the determination made by Evans *et al.*[122] (direct, 5.96 eV). The exciton binding energy quoted by Watanabe *et al.*[141] (0.149 eV, Fig. 2.10a) was far less than that calculated by Arnaud *et al.*[142] using the GW approximation (0.72 eV). Their calculations showed that low-lying excitons are tightly confined within the layers and, as such, h-BN has an indirect band gap of 5.95 eV. GW and dispersion corrected DFT calculations by Gao[143] confirmed this indirect band gap at 5.84 eV. DFT calculations also employing the GW approximation by Paleari *et al.*[144] show that the exciton binding energy was indeed 0.7 eV for bulk h-BN, but 1.93 eV for monolayer and 1.32 eV for pentalayer h-BN. They observed the band gap to decrease from 7.3 eV as a monolayer to 6.5 eV for pentalayer and 6.25 eV for bulk h-BN (indirect). A tight binding model was proposed to describe the splitting of the excitonic states in the multilayer systems. The predictions agreed with observations of excitonic emission made using luminescence spectroscopy for h-BN down to six layers.[145] However, monolayer h-BN produces two deep UV lines as a doublet in its PL spectrum, with valley physics becoming important in order to assign them, as observed by Elias *et al.*[146] for epitaxially grown h-BN by molecular beam epitaxy (MBE), as shown in Fig. 2.10b. These lines indicate an indirect-direct band gap crossover in monolayer h-BN and agree with the large exciton binding energy calculated by Paleari *et al.*[144] of 1.93 eV (Fig. 2.11). Thus, for few-layer and bulk h-BN, an indirect band gap exists, though for monolayer h-BN, there is a direct band gap. Thus, monolayer h-BN holds particular relevance for future photonic applications.[147]

h-BN has two signature peaks in the IR spectrum; BN stretching at 1,390 cm^{-1} and B-N-B bending at 780 cm^{-1}.[148] Geick *et al.*[148] performed the first spectroscopic study of h-BN and found the TO and LO phonon frequencies to be 1,367 cm^{-1} and 828 cm^{-1}, respectively, differing markedly to those of c-BN. These results were in agreement with theoretical estimations by Ohba *et al.* reported many years later.[149] As with IR spectroscopy, Raman scattering can clearly identify the presence of h-BN with the E_{2g} mode (analogous to the G peak for graphene) first measured at 1,367 cm^{-1}, in good agreement with the theoretical prediction of Ohba *et al.*,[149] 1,382 cm^{-1}.

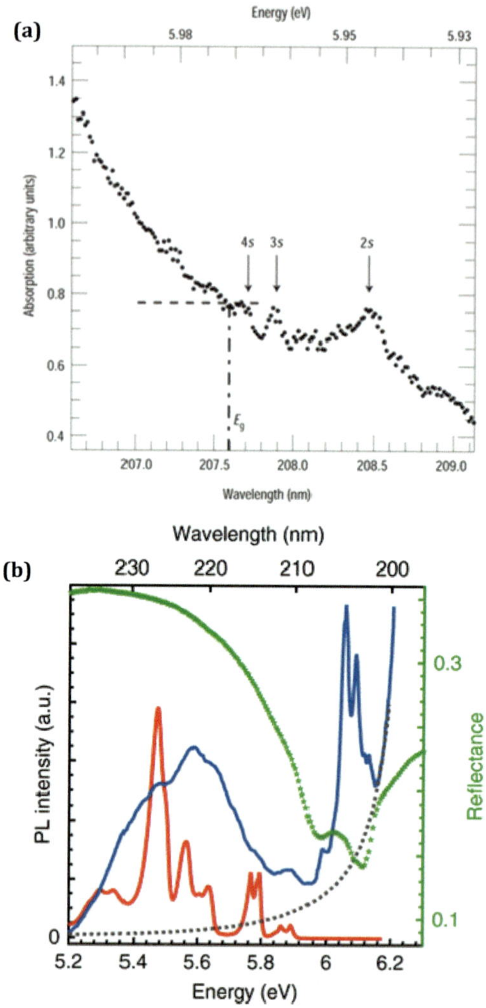

Figure 2.10 (a) Fundamental absorption spectra measured by Watanabe *et al.* showing key excitonic peaks, and the plateau indicating the direct band gap energy.[141] Reprinted by permission from Springer Nature Customer Service Center GmbH: Springer Nature. Watanabe, K. *et al.* Direct-band gap properties and evidence for ultraviolet lasing of hexagonal boron nitride single crystal. *Nat. Mater.* 2004, 3 (6), 404–409. Copyright 2004. (b) Reflectance spectrum (green), PL (blue) of monolayer h-BN on graphite at 10 K measured by Elias *et al.*, compared to bare graphite emission spectrum (gray) and PL of bulk h-BN (red).[146] Distributed under a Creative Commons CC BY license.

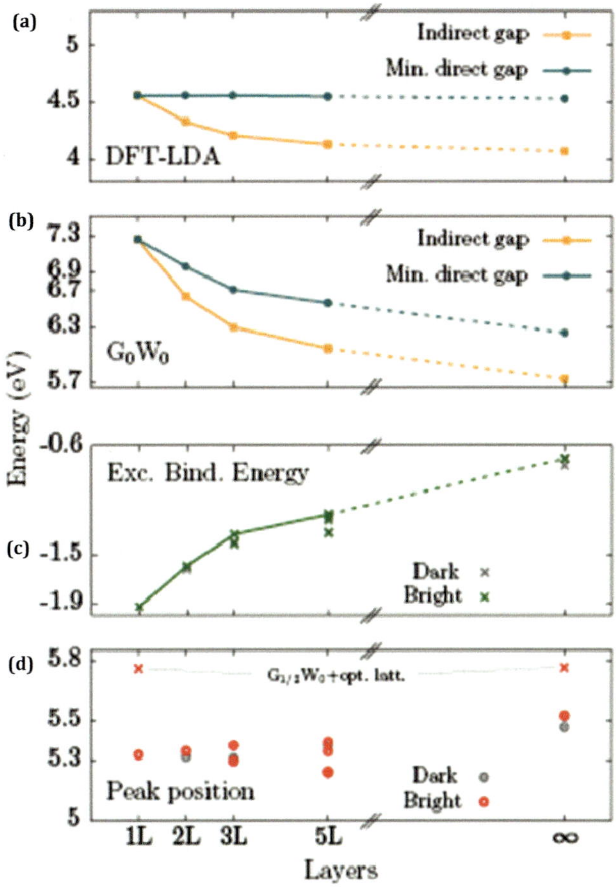

Figure 2.11 DFT band gaps calculated with local density approximation (LDA) (i) and G0W0 (ii) methods. (iii) Exciton binding energies of the lowest bound excitons and (iv) excitonic peak positions.[144] Paleari, F. *et al.* Excitons in few-layer hexagonal boron nitride: Davydov splitting and surface localization. *2D Mater.* 2018, 5 (4), 045017 © IOP Publishing. Reproduced with permission. All rights reserved.

Gorbachev *et al.*[150] made measurements of the peak position as a function of the number of layers and showed that bulk h-BN exhibits a peak at 1,366 cm^{-1} while films of 2, 4, 6, and 12 layers exhibited a slight red-shift from this bulk value (Fig. 2.12). Monolayer h-BN consistently showed a peak position around 1,368–1,370 cm^{-1} and importantly, a peak intensity 50 times lower than that of bulk h-BN. Therefore, Raman can be used to distinguish between one, two, or more BN layers in h-BN phases. Nemanich *et al.*[118] showed the peak position and broadening also depends on the h-BN domain size in polycrystalline samples, and extrapolated that the peak position for h-BN crystal should be 1,366 cm^{-1}.

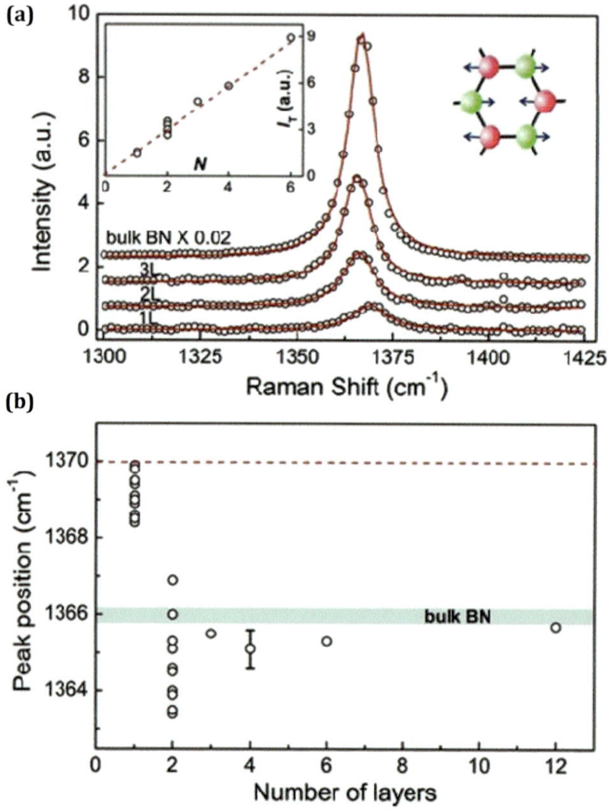

Figure 2.12 (a) Raman spectra measured by Gorbachev *et al.* for h-BN as a function of the number of layers, N. The left inset shows the integrated intensity and the right inset shows the phonon mode responsible in the BN ring.[150] (b) E_{2g} peak position as a function of the number of layers compared to bulk BN (shaded bar) and that predicted for monolayer h-BN (red dashed line).[150] Reprinted with permission from Gorbachev, R. V. *et al.* Hunting for monolayer boron nitride: Optical and Raman signatures. *Small* 2011, 7 (4), 465–468. Copyright 2011 John Wiley and Sons.

2.4.3 Optical Properties of BNNTs

The near-identical properties of h-BN and BNNT make it difficult to separate them or quantify their ratios in a mixture. Harrison *et al.*[151] demonstrate that IR peak ratios of the transverse and buckling modes (LO:TO) can be used to quantify the h-BN content in a BNNT sample. Regarding IR, BNNTs present approximately the same LO and TO as h-BN, although with an additional peak around 1,545 cm^{-1} arising from the vibration mode along the tangential direction of the BNNT.[152] Alon[153] found that armchair BNNTs have 9 Raman- and 14 IR-active phonon modes, ZZ BNNTs have 14 Raman and 8 IR-active modes and chiral BNNTs have 15 Raman- and 9 IR-active modes. Fakrach *et al.*[154] reported a slight dependence of these phonon modes and wavenumbers on chirality and diameter based on theoretical calculations, in agreement with experimental works.[155–157]

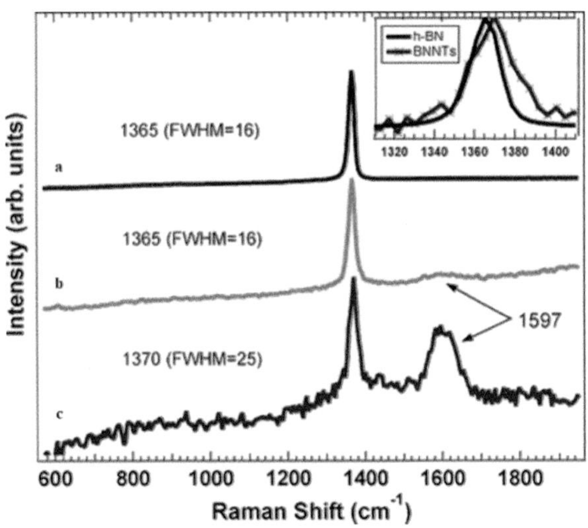

Figure 2.13 Raman spectra measured by Arenal, R. *et al.* excited at 229 nm on (a) BNNT, (b) h-BN, and (c) highly crystalline h-BN powder.[158] Reprinted with permission from Arenal, R. *et al.* Raman spectroscopy of single-wall boron nitride nanotubes. *Nano Lett.* 2006, 6 (8), 1812–1816. Copyright 2006 American Chemical Society.

Regarding Raman spectroscopy, the A_1 tangential mode of BNNT at 1,370 cm^{-1} was elucidated by Arenal *et al.*[158] for SWBNNTs as shown in Fig. 2.13. The authors explored the frequency change compared to bulk h-BN (1,365 cm^{-1}). The lower frequency for h-BN relates to the slightly lower lattice constant for monolayer h-BN vs. bulk h-BN. Neighboring h-BN sheets (or tube walls) cause softening of the phonons and elongation of the sp^2 bonds, causing the lower frequency for bulk h-BN. The presence of multiple walls can also see the broadening of the E_{2g} peak, masking the presence of the A_1 peak.[159] The presence and value of the A_1 peak agree with prior *ab initio* predictions by Wirtz *et al.*[160], which also showed the very weak A_{1g} peak just above 800 cm^{-1}, likely only detectable for very narrow tubes. The computed spectra reported by Aydin[161] also showed the A_{1g} peak, however, the A_1 and E_{2g} peaks appeared red-shifted in comparison 1,355 cm^{-1} and 1,330 cm^{-1}, respectively, and converged to 1,366 cm^{-1} and 1,360 cm^{-1} with increasing SWBNNT diameter.

Zhi *et al.* [162] observed the luminescence of BNNTs at 3.3 eV and 4.1 eV following electron beam excitation. Importantly, the emission of UV light was very stable during the experiment with no attenuation observed after 10 min of irradiation, reflecting the high stability of BNNTs. Multiple luminescence studies around this time observed different peaks for both MWBNNTs[163] and SWBNNTs.[164] Evidently (and unsurprisingly), the quality of the sample influences its luminescent properties. Nevertheless, luminescence was observed along the entire length of the tube. PL spectra of MWBNNTs were presented by Wu *et al.*,[159] and showed wide emission peaks at 3.5 eV and 4.2 eV, albeit much less intense than those observed for h-BN. Time-resolved PL spectroscopy enabled the elucidation of the recombination mechanism during emission, showing signs of typical band-edge luminescence behavior. Distinctively, PL emission was split into two regimes for MWBNNTs. In the first regime, recombination of electrons and holes following excitation is faster in BNNTs as the process is confined to single BN sheets/walls. In the second regime, recombination occurs between adjacent BN walls, thus taking much longer in BNNTs than h-BN, due to the indirect band gap. The intensity of the first regime far outweighs that of the second. Importantly, Jaffrenou *et al.*[165] showed that a bundle of SWBNNTs produced almost the same luminescence spectra as an individual tube (Fig. 2.14a). The broad, weak band around 4 eV

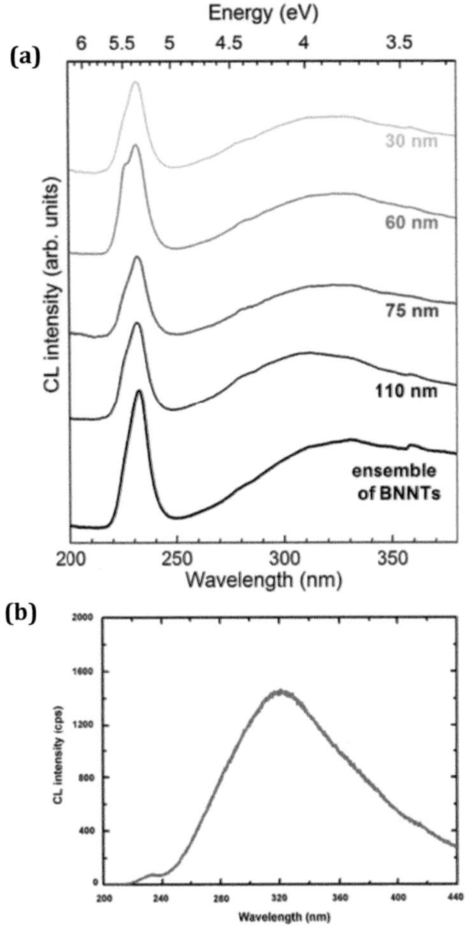

Figure 2.14 (a) Cathodoluminescence spectra measured by Jaffrennou *et al.* for individual MWBNNTs with varying outer diameters and a BNNT bundle at 100 K.[165] Reprinted with permission from Jaffrennou, P. *et al.* Near-band-edge recombination in multiwalled boron nitride nanotubes: Cathodoluminescence and photoluminescence spectroscopy measurements. *Phys. Rev. B* 2008, 77 (23), 235422. Copyright 2008 by the American Physical Society. (b) Cathodoluminescence spectra measured at room temperature by Huang *et al.* for boron oxide chemical vapor deposition (BOCVD)-synthesized BNNTs.[168] Huang, Y. *et al.* Bulk synthesis, growth mechanism, and properties of highly pure ultrafine boron nitride nanotubes with diameters of sub-10 nm. *Nanotechnology* 2011, 22 (14), 145602 © IOP Publishing. Reproduced with permission. All rights reserved.

(320 nm) was attributed to impurity and defect centers and confirmed *via* PL investigation across a range of temperatures. UV luminescence was observed with two peaks around 5.4 eV attributed to near-band gap bound-excitonic recombination, an exciting prospect for the use of BNNTs in optoelectronic devices. The other peak was previously attributed to free exciton recombination.[141] In contrast to the previously observed peaks showing uniform light spatial distribution,[164] the luminescence for the high energy peaks was quenched at the interface with the amorphous carbon membrane used during the luminescence measurement, a phenomenon observed previously with semiconducting CNTs in contact with metallic CNTs.[166] These deep-UV peaks are typically only observed at room temperature with high-quality materials such as the pure h-BN crystal[141] and carefully prepared BNNTs.[167] Notably, Huang *et al.* observed this luminescence at room temperature for CVD-grown BNNTs, as shown in Fig. 2.14b.[168]

Shortly after their initial proposal,[169] Blasé *et al.* showed through LDA calculations that, unlike CNTs, BNNTs exhibit the same band gap (~5.5 eV) regardless of their diameter or chirality due to localized charge density within the tube.[170] DFT calculations reported by Chou *et al.*[171] similarly showed that narrow BNNTs have smaller band gaps (e.g., (6,0) ~ 3 eV, (12,0) ~4.6 eV), while the band gap converged to 5.03 eV for ZZ tubes of diameter (20,0) and above. For armchair tubes, this convergence to 5.03 eV occurred for much narrower tubes, around (7,0); in either case, however, tubes of this size are not typically observed experimentally. Interestingly, slight changes in the band gap were observed between BNNTs of 100 nm and 2.5 nm diameter, *via* PL spectroscopy.[172] The band gap of BNNTs was shown to be tunable by such external influences as doping[173–176] and by applying an electric field.[177] Lauret *et al.*[178] reported the first experiments on optical transitions in SWBNNTs at room temperature, showing two low-energy absorption lines at 4.45 eV and 5.5 eV, the former notably lower than the band gap of h-BN. Band gap measurements *via* EELS shortly thereafter by Arenal *et al.*[179] revealed a geometry-independent BNNT band gap of 5.8 eV (Fig. 2.15).

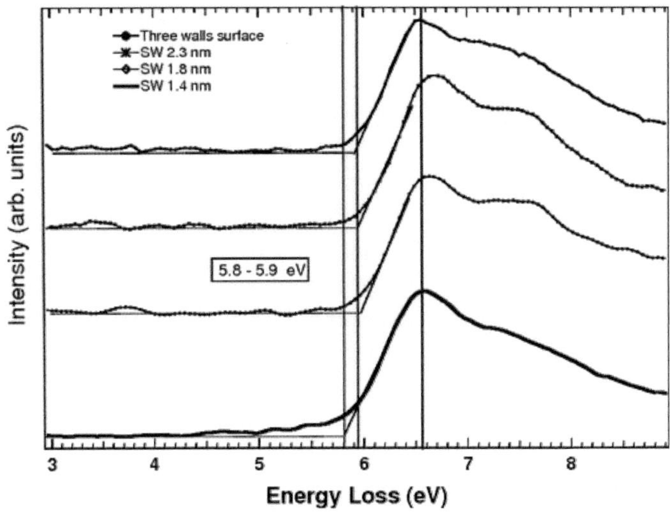

Figure 2.15 EELS spectra measured by Arenal, R. *et al.* showing the band gap of SWBNNTs with varying diameters as well as a three-walled BNNT.[179] Reprinted with permission from Arenal, R. *et al.* Electron energy loss spectroscopy measurement of the optical gaps on individual boron nitride single-walled and multi-walled nanotubes. *Phys. Rev. Lett.* 2005, 95 (12), 127601. Copyright 2005 by the American Physical Society.

The observed "optical gap" between absorptions for BNNTs vs. h-BN was attributed to the sensitive dependency on the first excitonic peak to the structure of the tube, though it converges quickly to that of h-BN.[180] The first, the much weaker absorption peak at 4.45 eV may have been a result of external perturbations like surrounding BNNTs; indeed, theory predicts for a single (8,0) BNNT that the lowest energy of the first exciton (K) is 4.63 eV,[181] though it was also observed as two peaks at 4.44 eV and 4.52 eV by Chen *et al.*[163] *via* PL measurements. The latter study determined the direct band gap to be 5.75 eV.

2.5 Electronic Properties

In terms of pure BN materials, their electronic properties do not differ between phases; they are intrinsically insulators. As we discuss in Chapter 4, the use of BN materials in electrical applications revolves around this intrinsic property, or is based on doping or the formation of a heterostructure with one or more highly conductive materials such as graphene and transition metal dichalcogenides (TMDs). We provide key examples regarding the properties of some of these materials; our focus here is the properties of pure BN materials, which are summarized in Table 2.4. We point the interested reader to important reviews such as those by Zhang,[184] Yankowitz *et al.*,[185] Kim *et al.*[186], and Wang *et al.*[187] Similarly, Lee *et al.*[188] discuss the functionalization of BNNTs. Numerous experimental techniques have been used for doping of BN materials such as sputtering,[189] ion implantation,[190] high-pressure high-temperature synthesis,[191, 192] CVD[193, 194], and substitution reactions.[195] The purpose of doping is to increase the mobility of electrons (i.e., the conductivity) within a material. By *p*-type doping, a defect or impurity is introduced in the form of an atom with fewer valence electrons than an atom that is replaced. This creates a hole in the valence band and electrons within the valence band become more mobile. For *n*-type doping, more electrons are available and can carry a charge, allowing more movement over the band gap in materials where otherwise the gap would be too large. Significant research has been performed on the properties of h-BN heterostructures, which we will discuss below. While less common, BNNTs have also been investigated for use as a dielectric material in composite materials.

Table 2.4 Electronic properties of BN phases

Electronic property	c-BN	h-BN	BNNT
Conductivity ($S\ cm^{-1}$)	6×10^{-8} [202]	10^{-8} [189]	
Resistivity (Ω cm)	10^{11} [203]	10^{11} [212]	300 [247]
Dielectric Constant	7.1 [59]	2–4 [220]	5.90 [257]
Breakdown Field ($MV\ cm^{-1}$)	6 [258] 8 [59]	12 [259]	12 [238]

2.5.1 Electronic Properties of c-BN

Unlike diamond, c-BN can be doped easily with both *p*- and *n*-type impurities,[196, 197] intentionally and unintentionally.[191, 194, 198, 199]

Gubanov *et al.*[200] showed this using the full potential and tight binding methods. Both B and N defect states appear to form well-defined narrow bands. B defect states are located near the top of the valence band and are partially occupied. For N, the vacancies overlap with states at the edge of the conduction band. DFT calculations by Orellana and Chacham[201] showed that B and N vacancies are the most stable defects in nonstoichiometric c-BN for *p*-type and *n*-type conditions, respectively. For example, S-doping of c-BN increased charge carrier concentrations at room temperature by two orders of magnitude.[191] Small concentrations of Si-doping in c-BN significantly lower resistivity,[202–204] as the Si atoms bond to N and replace B.[205, 206] *p*-type dopants such as Be,[207] Mg[208], and Zn[189] have successfully increased both carrier mobility and conductivity. Zhang extensively reviews the methods and results of doping c-BN.[184] Figure 2.16 shows an example of Si-doped c-BN films that exhibit enhanced conductivity with increasing Si concentration, while Fig. 2.17 shows current-voltage curves for h-BN and F-doped h-BN.

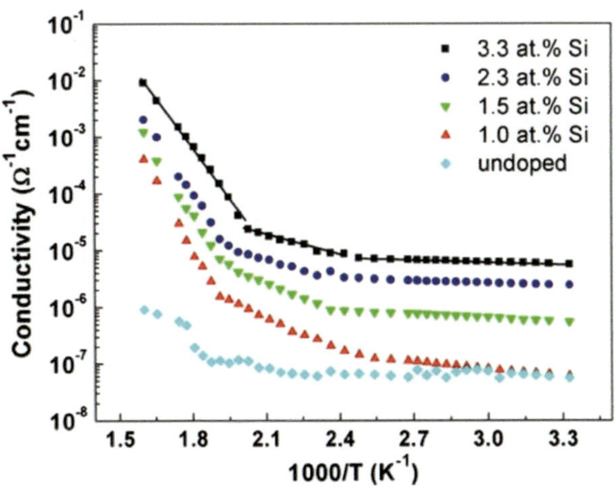

Figure 2.16 Conductivity of c-BN films doped with Si as a function of temperature.[202] Reprinted from Ying, J. *et al.* Electrical transport properties of the Si-doped cubic boron nitride thin films prepared by *in situ* co-sputtering. *J. Appl. Phys.* 2011, 109 (2), 023716., with the permission of AIP Publishing.

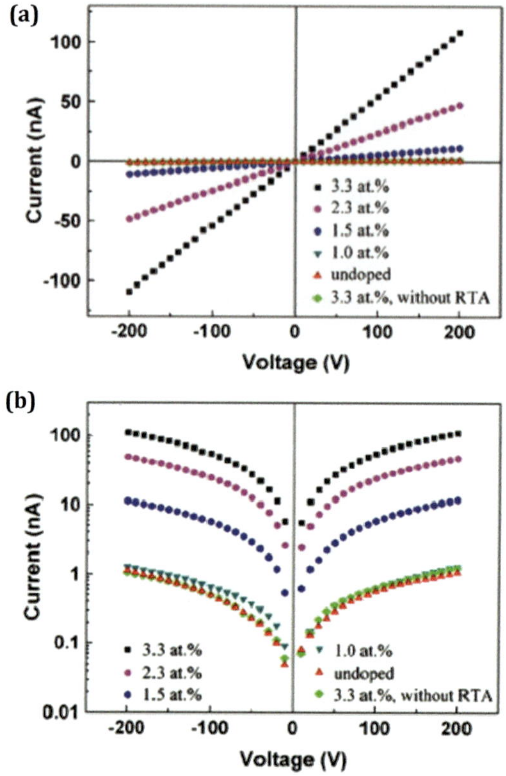

Figure 2.17 Current-voltage (I–V) curves measured by Ying, J. *et al.* for Si-doped c-BN on (a) linear and (b) logarithmic scales. [205] Reprinted from Ying, J. *et al.* Effects of silicon incorporation on composition, structure and electric conductivity of cubic boron nitride thin films. *Diamond Relat. Mater.* 2010, 19 (11), 1371–1376. Copyright 2010, with permission from Elsevier.

2.5.2 Electronic Properties of h-BN

As discussed in Section 2.4.2, monolayer h-BN has a direct band gap of around 5.95 eV.[141] The insulating properties of h-BN make it an ideal dielectric, with a favorable dielectric constant and breakdown field (Table 2.4). Notably, these properties can be altered by strain[209] though are only weakly influenced by layer thickness.[210] Early measurements of the resistivity of polycrystalline h-BN by Carpenter and Kirby[211] at temperatures between 700 °C and 1,400 °C showed high resistivity values increase from 10^4 Ω m to 10^{12} Ω m over this range. Room temperature resistivity was measured by Uddin *et al.*[212] to be approximately 10^{11} Ω cm. Unlike c-BN, there have been a few reports of h-BN doping, though *p*-type doping following Be ion implantation has been seen to increase conductivity[189] and charge carrier mobility[213, 214], and lower resistivity significantly to approximately 12 Ω cm at room temperature. Fluorination of h-BN also marked increases in conductivity.[215]

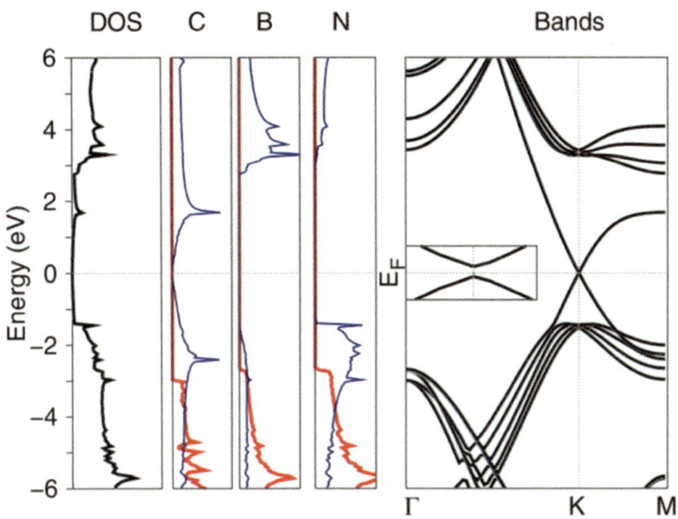

Figure 2.18 Band structure along with the ΓK and KM directions in reciprocal space and the total and projected densities of states (DOSs) published by Giovannetti, G. *et al.* for graphene on h-BN. Inset is zoomed in on the bands around the K point revealing an opening of the band gap.[218] Reprinted with permission from Giovannetti, G. *et al.* Substrate-induced band gap in graphene on hexagonal boron nitride: Ab initio density functional calculations. *Phys. Rev. B* 2007, 76 (7), 073103. Copyright 2007 by the American Physical Society.

Giovannetti *et al.*[218] predicted that stacking graphene and h-BN (made simple by their small 1.7% lattice mismatch) will open a band gap in the graphene sublattice of 53 meV (Fig. 2.18). This was followed by Dean *et al.*,[219] who succeeded in producing graphene on h-BN *via* mechanical transfer and reported mobilities of 60,000 $cm^2\ V^{-1}\ s^{-1}$; three times larger than that using SiO_2. Moreover, the chemical stability of the h-BN layer effectively protected graphene from chemical degradation. This led to an explosion of interest in h-BN/graphene heterostructures[185–187, 220–225] as well as h-BN/TMD heterostructures.[226–229] Significant research effort has been made to investigate the phenomenon of dielectric screening in these materials.[230–235] We discuss the application of h-BN in these heterostructures further in Chapter 4.

2.5.3 Electronic Properties of BNNTs

Theoretical work by Blase *et al.*[170] shortly after the discovery of BNNTs showed that BNNTs exhibited electronic properties independent of their chirality and diameter; a striking difference to CNTs.[236] Much like h-BN, BNNTs, and bundles of BNNTs are insulators.[45] Carrier mobilities of pure BNNTs are low at approximately 0.35 $cm^2\ V^{-1}\ s^{-1}$.[237] BNNTs can withstand voltages of 10 V and above with no degradation to the material.[238] Doping BNNTs with C saw a reduction in the band gap to 2 eV,[239, 240] an effect also seen with O doping.[241] Further attempts were made to dope with C and enhance BNNT electronic properties, though characteristics of these doped-BNNTs remained independent of chirality, diameter, and even tube quality.[242–246] Tang *et al.*[247] synthesized F-doped BNNTs and saw a reduction in the resistivity by three orders of magnitude, compared to undoped BNNTs (300 Ω cm to 0.2–0.6 Ω cm, respectively). Like doping with C,[246] F-doped BNNTs become *p*-type semiconductors[247–250] and additionally exhibit ferromagnetic properties.[251] Figure 2.19 shows the enhanced conductivity of BNNTs with fluorination. Electrically, the use of BNNTs is slated for that of an insulating, protective layer surrounding a CNT, predicted to be extremely useful electronic materials.[252, 253] These structures have been observed over two decades ago accidentally,[254] and experimental synthesis attempts have largely yielded coaxial nanotube structures with poorly refined structure.[255] Recently, however, Xiang *et* al.[256] have synthesized and characterized monocrystalline 1D van der Waals heterostructures comprising an inner CNT, middle BNNT, and outer MoS_2 nanotube. The insulating BNNT did not influence the intrinsic electronic transport of the inner CNT, yet due to the higher thermal stability of the

BNNT, the heteronanotubes were protected against oxidation and thermal degradation at temperatures around 700 °C rather than 400 °C for pure CNTs. We discuss the use of BNNTs as such insulating and protective materials further in Chapter 4.

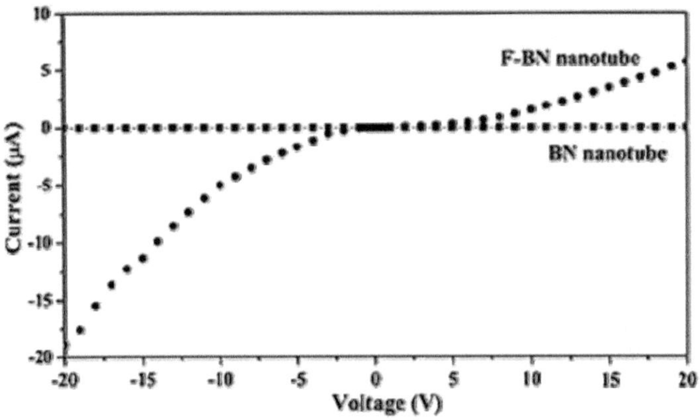

Figure 2.19 Current-voltage (I-V) curve measured by Tang, C. *et al.* comparing a defect-free BNNT and one functionalized with F.[247] Reprinted with permission from Tang, C. *et al.* Fluorination and electrical conductivity of BN nanotubes. *J. Am. Chem. Soc.* 2005, 127 (18), 6552–6553. Copyright 2005 American Chemical Society.

References

1. Kushch, V. I.; Dub, S. N.; Litvin, P. M., Determination of the Young modulus from elastic section of the Berkovich indenter loading curve. *J. Superhard Mater.* **2007**, *29* (4), 228–234.
2. Newey, D.; Wilkins, M. A.; Pollock, H. M., An ultra-low-load penetration hardness tester. *J. Phys. E.* **1982**, *15* (1), 119–122.
3. Pethicai, J. B.; Hutchings, R.; Oliver, W. C., Hardness measurement at penetration depths as small as 20 nm. *Philos. Mag. A.* **1983**, *48* (4), 593–606.
4. Wentorf, R. H., Cubic form of boron nitride. *J. Chem. Phys.* **1957**, *26* (4), 956–956.
5. Akio, H.; Shuji, Y., Sintered compact for a machining tool and a method of producing the compact. **1982**.
6. Bovenkerk, H.; Malloy, G., Method of making boron-alloyed diamond compacts and beryllium-alloyed cubic boron nitride compacts. **1971**.

7. Nakai, T.; Goto, M., Hard sintered compact for tools. **1990**.
8. Nakai, T.; Kono, Y.; Yazu, S.; Akio, H., Sintered compact for use in a tool and the method for producing the same. **1983**.
9. Wentorf, R. H., Abrasive material and preparation thereof. **1958**.
10. Wentorf, R. H.; De Lai, A., Cubic boron nitride compact and method for its production. **1964**.
11. Corrigan, F. R., Method for producing large cubic boron nitride particles. **1978**.
12. Sirota, N. N., Method for converting hexagonal boron nitride to a new structure. **1973**.
13. Mirkarimi, P. B.; Medlin, D. L.; McCarty, K. F.; Dibble, D. C.; Clift, W. M.; Knapp, J. A.; Barbour, J. C., The synthesis, characterization, and mechanical properties of thick, ultrahard cubic boron nitride films deposited by ion-assisted sputtering. *J. Appl. Phys.* **1997**, *82* (4), 1617–1625.
14. Hainsworth, S. V.; Chandler, H. W.; Page, T. F., Analysis of nanoindentation load-displacement loading curves. *J. Mater. Res.* **1996**, *11* (8), 1987–1995.
15. Hainsworth, S. V.; McGurk, M. R.; Page, T. F., The effect of coating cracking on the indentation response of thin hard-coated systems. *Surf. Coat. Technol.* **1998**, *102* (1), 97–107.
16. Shtansky, D. V.; Yamada-Takamura, Y.; Yoshida, T.; Ikuhara, Y., Mechanism of nucleation and growth of cubic boron nitride thin films. *Sci. Technol. Adv. Mater.* **2000**, *1* (4), 219–225.
17. Mei, H. Y.; Cai, X. H.; Tang, M.; Hui, Q.; Song, Q.; Wang, M., Electronic and mechanic properties of a new cubic boron nitride. *Comput. Mater. Sci.* **2019**, *162*, 111–115.
18. Mei, H. Y.; Pang, Y.; Liu, D. Y.; Cheng, N.; Zheng, S.; Song, Q.; Wang, M., Electronic and mechanic properties of trigonal boron nitride by first-principles calculations. *Physica E.* **2018**, *101*, 16–21.
19. Sumiya, H.; Uesaka, S.; Satoh, S., Mechanical properties of high purity polycrystalline cBN synthesized by direct conversion sintering method. *J. Mater. Sci.* **2000**, *35* (5), 1181–1186.
20. Hall, E. O., The deformation and ageing of mild steel: III discussion of results. *Proc. Phys. Soc.* **1951**, *64* (9), 747–753.
21. Petch, N. J., The Cleavage Strength of Polycrystals. *Tetsu-To-Hagane.* **1953**, *174*.

22. Dub, S. N.; Petrusha, I. A., Mechanical properties of polycrystalline cBN obtained from pyrolytic gBN by direct transformation technique. *High Press. Res.* **2006**, *26* (2), 71–77.

23. Dubrovinskaia, N.; Solozhenko, V. L.; Miyajima, N.; Dmitriev, V.; Kurakevych, O. O.; Dubrovinsky, L., Superhard nanocomposite of dense polymorphs of boron nitride: Noncarbon material has reached diamond hardness. *Appl. Phys. Lett.* **2007**, *90* (10), 101912.

24. Lu, L.; Chen, X.; Huang, X.; Lu, K., Revealing the maximum strength in nanotwinned copper. *Science.* **2009**, *323* (5914), 607.

25. Pande, C. S.; Cooper, K. P., Nanomechanics of Hall–Petch relationship in nanocrystalline materials. *Prog. Mater Sci.* **2009**, *54* (6), 689–706.

26. Schiøtz, J.; Di Tolla, F. D.; Jacobsen, K. W., Softening of nanocrystalline metals at very small grain sizes. *Nature.* **1998**, *391* (6667), 561–563.

27. Schiøtz, J.; Jacobsen, K. W., A maximum in the strength of nanocrystalline copper. *Science.* **2003**, *301* (5638), 1357.

28. Solozhenko, V. L.; Kurakevych, O. O.; Le Godec, Y., Creation of nanostuctures by extreme conditions: High-pressure synthesis of ultrahard nanocrystalline cubic boron nitride. *Adv. Mater.* **2012**, *24* (12), 1540–1544.

29. Tian, Y.; Xu, B.; Yu, D.; Ma, Y.; Wang, Y.; Jiang, Y.; Hu, W.; Tang, C.; Gao, Y.; Luo, K.; Zhao, Z.; Wang, L.-M.; Wen, B.; He, J.; Liu, Z., Ultrahard nanotwinned cubic boron nitride. *Nature.* **2013**, *493* (7432), 385–388.

30. Solozhenko, V. L.; Bushlya, V.; Zhou, J., Mechanical properties of ultra-hard nanocrystalline cubic boron nitride. *J. Appl. Phys.* **2019**, *126* (7), 075107.

31. Lee, C.; Wei, X.; Kysar, J. W.; Hone, J., Measurement of the elastic properties and intrinsic strength of monolayer graphene. *Science.* **2008**, *321* (5887), 385.

32. Han, T.; Luo, Y.; Wang, C., Effects of temperature and strain rate on the mechanical properties of hexagonal boron nitride nanosheets. *J. Phys. D: Appl. Phys.* **2013**, *47* (2), 025303.

33. Bosak, A.; Serrano, J.; Krisch, M.; Watanabe, K.; Taniguchi, T.; Kanda, H., Elasticity of hexagonal boron nitride: Inelastic X-ray scattering measurements. *Phys. Rev. B.* **2006**, *73* (4), 041402.

34. Wu, J.; Wang, B.; Wei, Y.; Yang, R.; Dresselhaus, M., Mechanics and mechanically tunable band gap in single-layer hexagonal boron-nitride. *Mater. Res. Lett.* **2013**, *1* (4), 200–206.

35. Peng, Q.; Ji, W.; De, S., Mechanical properties of the hexagonal boron nitride monolayer: Ab initio study. *Comput. Mater. Sci.* **2012**, *56*, 11–17.

36. Mirnezhad, M.; Ansari, R.; Rouhi, H., Mechanical properties of multilayer boron nitride with different stacking orders. *Superlattices Microstruct.* **2013**, *53*, 223–231.

37. Thomas, S.; Ajith, K. M.; Valsakumar, M. C., Directional anisotropy, finite size effect and elastic properties of hexagonal boron nitride. *J. Phys.: Condens. Matter.* **2016**, *28* (29), 295302.

38. Song, L.; Ci, L.; Lu, H.; Sorokin, P. B.; Jin, C.; Ni, J.; Kvashnin, A. G.; Kvashnin, D. G.; Lou, J.; Yakobson, B. I.; Ajayan, P. M., Large scale growth and characterization of atomic hexagonal boron nitride layers. *Nano Lett.* **2010**, *10* (8), 3209–3215.

39. Kim, S. M.; Hsu, A.; Park, M. H.; Chae, S. H.; Yun, S. J.; Lee, J. S.; Cho, D.-H.; Fang, W.; Lee, C.; Palacios, T.; Dresselhaus, M.; Kim, K. K.; Lee, Y. H.; Kong, J., Synthesis of large-area multilayer hexagonal boron nitride for high material performance. *Nat. Commun.* **2015**, *6* (1), 8662.

40. Falin, A.; Cai, Q.; Santos, E. J. G.; Scullion, D.; Qian, D.; Zhang, R.; Yang, Z.; Huang, S.; Watanabe, K.; Taniguchi, T.; Barnett, M. R.; Chen, Y.; Ruoff, R. S.; Li, L. H., Mechanical properties of atomically thin boron nitride and the role of interlayer interactions. *Nat. Commun.* **2017**, *8* (1), 15815.

41. Li, C.; Bando, Y.; Zhi, C.; Huang, Y.; Golberg, D., Thickness-dependent bending modulus of hexagonal boron nitride nanosheets. *Nanotechnology.* **2009**, *20* (38), 385707.

42. Chopra, N. G.; Zettl, A., Measurement of the elastic modulus of a multi-wall boron nitride nanotube. *Solid State Commun.* **1998**, *105* (5), 297–300.

43. Suryavanshi, A. P.; Yu, M.-F.; Wen, J.; Tang, C.; Bando, Y., Elastic modulus and resonance behavior of boron nitride nanotubes. *Appl. Phys. Lett.* **2004**, *84* (14), 2527–2529.

44. Golberg, D.; Costa, P. M. F. J.; Lourie, O.; Mitome, M.; Bai, X.; Kurashima, K.; Zhi, C.; Tang, C.; Bando, Y., Direct force measurements and kinking under elastic deformation of individual multiwalled boron nitride nanotubes. *Nano Lett.* **2007**, *7* (7), 2146–2151.

45. Arenal, R.; Wang, M.-S.; Xu, Z.; Loiseau, A.; Golberg, D., Young modulus, mechanical and electrical properties of isolated individual and bundled single-walled boron nitride nanotubes. *Nanotechnology.* **2011**, *22* (26), 265704.

46. Huang, Y.; Wu, J.; Hwang, K. C., Thickness of graphene and single-walled carbon nanotubes. *Phys. Rev. B.* **2006**, *74* (24), 245413.

47. Tanur, A. E.; Wang, J.; Reddy, A. L. M.; Lamont, D. N.; Yap, Y. K.; Walker, G. C., Diameter-dependent bending modulus of individual multiwall boron nitride nanotubes. *J. Phys. Chem. B.* **2013**, *117* (16), 4618–4625.

48. Verma, V.; Jindal, V. K.; Dharamvir, K., Elastic moduli of a boron nitride nanotube. *Nanotechnology.* **2007**, *18* (43), 435711.

49. Hernández, E.; Goze, C.; Bernier, P.; Rubio, A., Elastic properties of C and BxCyNz composite nanotubes. *Phys. Rev. Lett.* **1998**, *80* (20), 4502–4505.

50. Gardinier, C. F., *Ceram. Bull.* **1988**, *67* (6).

51. Monteiro, S. N.; Skury, A. L. D.; de Azevedo, M. G.; Bobrovnitchii, G. S., Cubic boron nitride competing with diamond as a superhard engineering material—An overview. *J. Mater. Res. Technol.* **2013**, *2* (1), 68–74.

52. Knittle, E.; Kaner, R. B.; Jeanloz, R.; Cohen, M. L., High-pressure synthesis, characterization, and equation of state of cubic C-BN solid solutions. *Phys. Rev. B.* **1995**, *51* (18), 12149–12156.

53. Gielisse, P. J.; Mitra, S. S.; Plendl, J. N.; Griffis, R. D.; Mansur, L. C.; Marshall, R.; Pascoe, E. A., Lattice infrared spectra of boron nitride and boron monophosphide. *Phys. Rev.* **1967**, *155* (3), 1039–1046.

54. Wakatsuki, M.; Ichinose, K.; Aoki, T., Synthesis of polycrystalline cubic BN. *Mater. Res. Bull.* **1972**, *7* (9), 999–1003.

55. Corrigan, F. R.; Bundy, F. P., Direct transitions among the allotropic forms of boron nitride at high pressures and temperatures. *J. Chem. Phys.* **1975**, *63* (9), 3812–3820.

56. Taniguchi, T.; Akaishi, M.; Yamaoka, S., Mechanical properties of polycrystalline translucent cubic boron nitride as characterized by the vickers indentation method. *J. Am. Ceram. Soc.* **2005**, *79*, 547–549.

57. Ichinose, K.; Wakatsuki, M.; Aoki, T.; Maeda, Y., *Rev. Phys. Chem. Jpn.* **1975**.

58. Hirano, S.-I.; Yamaguchi, T.; Naka, S., Effects of AIN additions and atmosphere on the synthesis of cubic boron nitride. *J. Am. Ceram. Soc.* **1981**, *64* (12), 734–736.

59. Samantaray, C. B.; Singh, R. N., Review of synthesis and properties of cubic boron nitride (c-BN) thin films. *Int. Mater. Rev.* **2005**, *50* (6), 313–344.

60. Solozhenko, V. L.; Dub, S. N.; Novikov, N. V., Mechanical properties of cubic BC2N, a new superhard phase. *Diamond Relat. Mater.* **2001**, *10* (12), 2228–2231.

61. Topsakal, M.; Cahangirov, S.; Ciraci, S., The response of mechanical and electronic properties of graphane to the elastic strain. *Appl. Phys. Lett.* **2010**, *96* (9), 091912.

62. Tang, D.-M.; Ren, C.-L.; Wei, X.; Wang, M.-S.; Liu, C.; Bando, Y.; Golberg, D., Mechanical properties of bamboo-like boron nitride nanotubes by *in situ* TEM and MD simulations: Strengthening effect of interlocked joint interfaces. *ACS Nano.* **2011**, *5* (9), 7362–7368.

63. Guhados, G.; Wan, W.; Sun, X.; Hutter, J. L., Simultaneous measurement of Young's and shear moduli of multi-walled carbon nanotubes using atomic force microscopy. *J. Appl. Phys.* **2007**, *101* (3), 033514.

64. Ghassemi, H. M.; Lee, C. H.; Yap, Y. K.; Yassar, R. S., Real-time fracture detection of individual boron nitride nanotubes in severe cyclic deformation processes. *J. Appl. Phys.* **2010**, *108* (2), 024314.

65. Kochaev, A., Elastic properties of noncarbon nanotubes as compared to carbon nanotubes. *Phys. Rev. B.* **2017**, *96* (15), 155428.

66. Natsuki, T.; Natsuki, J., Prediction of mechanical properties for hexagonal boron nitride nanosheets using molecular mechanics model. *Appl. Phys. A* **2017**, *123* (4), 283.

67. Zhao, Y.; Chen, X.; Park, C.; Fay, C. C.; Stupkiewicz, S.; Ke, C., Mechanical deformations of boron nitride nanotubes in crossed junctions. *J. Appl. Phys.* **2014**, *115* (16), 164305.

68. Wei, X.; Wang, M.-S.; Bando, Y.; Golberg, D., Tensile tests on individual multi-walled boron nitride nanotubes. *Adv. Mater.* **2010**, *22* (43), 4895–4899.

69. Kudin, K. N.; Scuseria, G. E.; Yakobson, B. I., C2F, BN, and C nanoshell elasticity from ab initio computations. *Phys. Rev. B.* **2001**, *64* (23), 235406.

70. Duclaux, L.; Nysten, B.; Issi, J. P.; Moore, A. W., Structure and low-temperature thermal conductivity of pyrolytic boron nitride. *Phys. Rev. B.* **1992**, *46* (6), 3362–3367.

71. Qu, W.; Bagchi, S.; Chen, X.; Chew, H. B.; Ke, C., Bending and interlayer shear moduli of ultrathin boron nitride nanosheet. *J. Phys. D: Appl. Phys.* **2019**, *52* (46), 465301.

72. Hamdi, I.; Meskini, N., Ab initio study of the structural, elastic, vibrational and thermodynamic properties of the hexagonal boron

nitride: Performance of LDA and GGA. *Physica B.* **2010**, *405* (13), 2785–2794.

73. Green, J. F.; Bolland, T. K.; Bolland, J. W., Theoretical elastic behavior for hexagonal boron nitride. *J. Chem. Phys.* **1976**, *64* (2), 656–662.
74. Zheng, M.; Chen, X.; Park, C.; Fay, C. C.; Pugno, N. M.; Ke, C., Nanomechanical cutting of boron nitride nanotubes by atomic force microscopy. *Nanotechnology.* **2013**, *24* (50), 505719.
75. Lee, S. M.; Cahill, D. G., Heat transport in thin dielectric films. *J. Appl. Phys.* **1997**, *81* (6), 2590–2595.
76. Sichel, E. K.; Miller, R. E.; Abrahams, M. S.; Buiocchi, C. J., Heat capacity and thermal conductivity of hexagonal pyrolytic boron nitride. *Phys. Rev. B.* **1976**, *13* (10), 4607–4611.
77. Norris, P. M.; Caffrey, A. P.; Stevens, R. J.; Klopf, J. M.; McLeskey, J. T.; Smith, A. N., Femtosecond pump–probe nondestructive examination of materials (invited). *Rev. Sci. Instrum.* **2003**, *74* (1), 400–406.
78. Balandin, A. A., Thermal properties of graphene and nanostructured carbon materials. *Nat. Mater.* **2011**, *10* (8), 569–581.
79. Balandin, A. A.; Ghosh, S.; Bao, W.; Calizo, I.; Teweldebrhan, D.; Miao, F.; Lau, C. N., Superior thermal conductivity of single-layer graphene. *Nano Lett.* **2008**, *8* (3), 902–907.
80. Calizo, I.; Balandin, A. A.; Bao, W.; Miao, F.; Lau, C. N., Temperature dependence of the raman spectra of graphene and graphene multilayers. *Nano Lett.* **2007**, *7* (9), 2645–2649.
81. Cai, W.; Moore, A. L.; Zhu, Y.; Li, X.; Chen, S.; Shi, L.; Ruoff, R. S., Thermal transport in suspended and supported monolayer graphene grown by chemical vapor deposition. *Nano Lett.* **2010**, *10* (5), 1645–1651.
82. Faugeras, C.; Faugeras, B.; Orlita, M.; Potemski, M.; Nair, R. R.; Geim, A. K., Thermal conductivity of graphene in corbino membrane geometry. *ACS Nano.* **2010**, *4* (4), 1889–1892.
83. Zhou, H.; Zhu, J.; Liu, Z.; Yan, Z.; Fan, X.; Lin, J.; Wang, G.; Yan, Q.; Yu, T.; Ajayan, P. M.; Tour, J. M., High thermal conductivity of suspended few-layer hexagonal boron nitride sheets. *Nano Res.* **2014**, *7* (8), 1232–1240.
84. Seol, J. H.; Jo, I.; Moore, A. L.; Lindsay, L.; Aitken, Z. H.; Pettes, M. T.; Li, X.; Yao, Z.; Huang, R.; Broido, D.; Mingo, N.; Ruoff, R. S.; Shi, L., Two-dimensional phonon transport in supported graphene. *Science.* **2010**, *328* (5975), 213.

85. Jo, I.; Pettes, M. T.; Kim, J.; Watanabe, K.; Taniguchi, T.; Yao, Z.; Shi, L., Thermal conductivity and phonon transport in suspended few-layer hexagonal boron nitride. *Nano Lett.* **2013**, *13* (2), 550–554.

86. Chang, C. W.; Fennimore, A. M.; Afanasiev, A.; Okawa, D.; Ikuno, T.; Garcia, H.; Li, D.; Majumdar, A.; Zettl, A., Isotope effect on the thermal conductivity of boron nitride nanotubes. *Phys. Rev. Lett.* **2006**, *97* (8), 085901.

87. Berman, R.; Simon, F. E.; Ziman, J. M., The thermal conductivity of diamond at low temperatures. *Proc. R. Soc. Lond. A.* **1953**, *220* (1141), 171–183.

88. Olson, J. R.; Pohl, R. O.; Vandersande, J. W.; Zoltan, A.; Anthony, T. R.; Banholzer, W. F., Thermal conductivity of diamond between 170 and 1200 K and the isotope effect. *Phys. Rev. B.* **1993**, *47* (22), 14850–14856.

89. Wei, L.; Kuo, P. K.; Thomas, R. L.; Anthony, T. R.; Banholzer, W. F., Thermal conductivity of isotopically modified single crystal diamond. *Phys. Rev. Lett.* **1993**, *70* (24), 3764–3767.

90. Novikov, N. V.; Podoba, A. P.; Perevertailo, V. M.; Shmegera, S. V.; Witek, A., Effect of isotope content on the cubic boron nitride lattice thermal conductivity. *Diamond Relat. Mater.* **2000**, *9* (3), 629–631.

91. Morelli, D. T.; Heremans, J. P.; Slack, G. A., Estimation of the isotope effect on the lattice thermal conductivity of group IV and group III-V semiconductors. *Phys. Rev. B.* **2002**, *66* (19), 195304.

92. Chen, K.; Song, B.; Ravichandran, N. K.; Zheng, Q.; Chen, X.; Lee, H.; Sun, H.; Li, S.; Udalamatta Gamage, G. A. G.; Tian, F.; Ding, Z.; Song, Q.; Rai, A.; Wu, H.; Koirala, P.; Schmidt, A. J.; Watanabe, K.; Lv, B.; Ren, Z.; Shi, L.; Cahill, D. G.; Taniguchi, T.; Broido, D.; Chen, G., Ultrahigh thermal conductivity in isotope-enriched cubic boron nitride. *Science.* **2020**, *367* (6477), 555.

93. Giri, A.; Hopkins, P. E., Achieving a better heat conductor. *Nat. Mater.* **2020**, *19* (5), 482–484.

94. Lindsay, L.; Broido, D. A., Enhanced thermal conductivity and isotope effect in single-layer hexagonal boron nitride. *Phys. Rev. B.* **2011**, *84* (15), 155421.

95. Cai, Q.; Scullion, D.; Gan, W.; Falin, A.; Zhang, S.; Watanabe, K.; Taniguchi, T.; Chen, Y.; Santos, E. J. G.; Li, L. H., High thermal conductivity of high-quality monolayer boron nitride and its thermal expansion. *Science Adv.* **2019**, *5* (6), eaav0129.

96. Mingo, N.; Yang, L., Phonon transport in nanowires coated with an amorphous material: An atomistic Green's function approach. *Phys. Rev. B.* **2003**, *68* (24), 245406.

97. Stewart, D. A.; Savić, I.; Mingo, N., First-principles calculation of the isotope effect on boron nitride nanotube thermal conductivity. *Nano Lett.* **2009**, *9* (1), 81–84.

98. Kim, P.; Shi, L.; Majumdar, A.; McEuen, P. L., Thermal transport measurements of individual multiwalled nanotubes. *Phys. Rev. Lett.* **2001**, *87* (21), 215502.

99. Jiang, J.-W.; Wang, J.-S., Theoretical study of thermal conductivity in single-walled boron nitride nanotubes. *Phys. Rev. B.* **2011**, *84* (8), 085439.

100. Belkerk, B. E.; Achour, A.; Zhang, D.; Sahli, S.; Djouadi, M. A.; Yap, Y. K., Thermal conductivity of vertically aligned boron nitride nanotubes. *App. Phys. Express.* **2016**, *9* (7), 075002.

101. Chen, Y.; Zou, J.; Campbell, S. J.; Le Caer, G., Boron nitride nanotubes: Pronounced resistance to oxidation. *Appl. Phys. Lett.* **2004**, *84* (13), 2430–2432.

102. Alam, M. T.; Bresnehan, M. S.; Robinson, J. A.; Haque, M. A., Thermal conductivity of ultra-thin chemical vapor deposited hexagonal boron nitride films. *Appl. Phys. Lett.* **2014**, *104* (1), 013113.

103. Yuan, C.; Li, J.; Lindsay, L.; Cherns, D.; Pomeroy, J. W.; Liu, S.; Edgar, J. H.; Kuball, M., Modulating the thermal conductivity in hexagonal boron nitride *via* controlled boron isotope concentration. *Communications Physics.* **2019**, *2* (1), 43.

104. Cai, Q.; Scullion, D.; Falin, A.; Watanabe, K.; Taniguchi, T.; Chen, Y.; Santos, E. J. G.; Li, L. H., Raman signature and phonon dispersion of atomically thin boron nitride. *Nanoscale.* **2017**, *9* (9), 3059–3067.

105. Chen, X.; Dmuchowski, C. M.; Park, C.; Fay, C. C.; Ke, C., Quantitative characterization of structural and mechanical properties of boron nitride nanotubes in high temperature environments. *Sci. Rep.* **2017**, *7* (1), 11388.

106. Wirtz, L.; Rubio, A., Optical and vibrational properties of boron nitride nanotubes. In *B-C-N Nanotubes and Related Nanomaterials*, 1 Ed.; Yoke Khin, Y. (ed.). Springer-Verlag New York: 2009; Vol. 1, XV, 299.

107. Brafman, O.; Lengyel, G.; Mitra, S. S.; Gielisse, P. J.; Plendl, J. N.; Mansur, L. C., Raman spectra of AℓN, cubic BN and BP. *Solid State Commun.* **1968**, *6* (8), 523–526.

108. Chrenko, R. M., Ultraviolet and infrared spectra of cubic boron nitride. *Solid State Commun.* **1974**, *14* (6), 511–515.

109. Eremets, M. I.; Gauthier, M.; Polian, A.; Chervin, J. C.; Besson, J. M.; Dubitskii, G. A.; Semenova, Y. Y., Optical properties of cubic boron nitride. *Phys. Rev. B.* **1995**, *52* (12), 8854–8863.

110. Cai, Y.; Zhang, L.; Zeng, Q.; Cheng, L.; Xu, Y., Infrared reflectance spectrum of BN calculated from first principles. *Solid State Commun.* **2007**, *141* (5), 262–266.

111. Ullmann, J.; Baglin, J. E. E.; Kellock, A. J., Effects of MeV ion irradiation of thin cubic boron nitride films. *J. Appl. Phys.* **1998**, *83* (6), 2980–2987.

112. Medlin, D. L.; Friedmann, T. A.; Mirkarimi, P. B.; Rez, P.; Mills, M. J.; McCarty, K. F., Microstructure of cubic boron nitride thin films grown by ion-assisted pulsed laser deposition. *J. Appl. Phys.* **1994**, *76* (1), 295–303.

113. Zhang, W. J.; Jiang, X.; Matsumoto, S., High-quality, faceted cubic boron nitride films grown by chemical vapor deposition. *Appl. Phys. Lett.* **2001**, *79* (27), 4530–4532.

114. Zhang, X. W.; Boyen, H. G.; Deyneka, N.; Ziemann, P.; Banhart, F.; Schreck, M., Epitaxy of cubic boron nitride on (001)-oriented diamond. *Nat. Mater.* **2003**, *2* (5), 312–315.

115. Friedmann, T. A.; Mirkarimi, P. B.; Medlin, D. L.; McCarty, K. F.; Klaus, E. J.; Boehme, D. R.; Johnsen, H. A.; Mills, M. J.; Ottesen, D. K.; Barbour, J. C., Ion-assisted pulsed laser deposition of cubic boron nitride films. *J. Appl. Phys.* **1994**, *76* (5), 3088–3101.

116. Werninghaus, T.; Hahn, J.; Richter, F.; Zahn, D. R. T., Raman spectroscopy investigation of size effects in cubic boron nitride. *Appl. Phys. Lett.* **1997**, *70* (8), 958–960.

117. Ben el Mekki, M.; Djouadi, M. A.; Guiot, E.; Mortet, V.; Pascallon, J.; Stambouli, V.; Bouchier, D.; Mestres, N.; Nouet, G., Structure investigation of BN films grown by ion-beam-assisted deposition by means of polarised IR and Raman spectroscopy. *Surf. Coat. Technol.* **1999**, *116–119*, 93–99.

118. Nemanich, R. J.; Solin, S. A.; Martin, R. M., Light scattering study of boron nitride microcrystals. *Phys. Rev. B.* **1981**, *23* (12), 6348–6356.

119. Yokoyama, H.; Okamoto, M.; Osaka, Y., Effects of a negative self-bias on the growth of cubic boron nitride prepared by plasma chemical vapor deposition. *Jpn. J. Appl. Phys.* **1991**, *30* (2R), 344.

120. Jäger, S.; Bewilogua, K.; Klages, C. P., Infrared spectroscopic investigations on h-BN and mixed h/c-BN thin films. *Thin Solid Films.* **1994**, *245* (1), 50–54.

121. Dovesi, R.; Pisani, C.; Roetti, C.; Dellarole, P., Exact-exchange Hartree-Fock calculations for periodic systems. IV. Ground-state properties of cubic boron nitride. *Phys. Rev. B.* **1981**, *24* (8), 4170–4176.

122. Evans, D. A.; McGlynn, A. G.; Towlson, B. M.; Gunn, M.; Jones, D.; Jenkins, T. E.; Winter, R.; Poolton, N. R. J., Determination of the optical band-gap energy of cubic and hexagonal boron nitride using luminescence excitation spectroscopy. *J. Phys.: Condens. Matter.* **2008**, *20* (7), 075233.

123. Xu, Y.-N.; Ching, W. Y., Calculation of ground-state and optical properties of boron nitrides in the hexagonal, cubic, and wurtzite structures. *Phys. Rev. B.* **1991**, *44* (15), 7787–7798.

124. Powers, M. J.; Benjamin, M. C.; Porter, L. M.; Nemanich, R. J.; Davis, R. F.; Cuomo, J. J.; Doll, G. L.; Harris, S. J., Observation of a negative electron affinity for boron nitride. *Appl. Phys. Lett.* **1995**, *67*, 3912.

125. Fomichev, V. A.; Rumsh, M. A., Investigation of X-ray spectra of hexagonal and cubic boron nitride. *J. Phys. Chem. Solids.* **1968**, *29* (6), 1015–1024.

126. Jiang, H.; Blaha, P., GW with linearized augmented plane waves extended by high-energy local orbitals. *Phys. Rev. B.* **2016**, *93* (11), 115203.

127. Gao, S.-P., Cubic, wurtzite, and 4H-BN band structures calculated using GW methods and maximally localized Wannier functions interpolation. *Comp. Mater. Sci.* **2012**, *61*, 266–269.

128. Klimeš, J.; Kaltak, M.; Kresse, G., Predictive GW calculations using plane waves and pseudopotentials. *Phys. Rev. B.* **2014**, *90* (7), 075125.

129. Tararan, A.; di Sabatino, S.; Gatti, M.; Taniguchi, T.; Watanabe, K.; Reining, L.; Tizei, L. H. G.; Kociak, M.; Zobelli, A., Optical gap and optically active intragap defects in cubic BN. *Phys. Rev. B.* **2018**, *98* (9), 094106.

130. McKenzie, D. R.; McFall, W. D.; Smith, H.; Higgins, B.; Boswell, R. W.; Durandet, A.; James, B. W.; Falconer, I. S., High pressure phases produced by low energy ion implantation with reference to cubic boron nitride. *Nucl. Instrum. Meth. B.* **1995**, *106* (1), 90–95.

131. Ren, S. L.; Rao, A. M.; Eklund, P. C.; Doll, G. L., Ellipsometric study of boron nitride thin-film growth on Si(100). *Appl. Phys. Lett.* **1993**, *62* (15), 1760–1762.

132. Schubert, M.; Rheinländer, B.; Franke, E.; Neumann, H.; Hahn, J.; Röder, M.; Richter, F., Anisotropy of boron nitride thin-film reflectivity spectra by generalized ellipsometry. *Appl. Phys. Lett.* **1997**, *70* (14), 1819–1821.

133. Cappellini, G.; Satta, G.; Palummo, M.; Onida, G., Optical properties of BN in cubic and layered hexagonal phases. *Phys. Rev. B.* **2001**, *64* (3), 035104.

134. Segura, A.; Artús, L.; Cuscó, R.; Taniguchi, T.; Cassabois, G.; Gil, B., Natural optical anisotropy of h-BN: Highest giant birefringence in a bulk crystal through the mid-infrared to ultraviolet range. *Physical Review Materials.* **2018**, *2* (2), 024001.

135. Lee, S.-Y.; Jeong, T.-Y.; Jung, S.; Yee, K.-J., Refractive index dispersion of hexagonal boron nitride in the visible and near-infrared. *Phys. Status Solidi B.* **2019**, *256* (6), 1800417.

136. Rand, M. J.; Roberts, J. F., Preparation and properties of thin film boron nitride. *J. Electrochem. Soc.* **1968**, *115* (4), 423–429.

137. Zupan, J.; Kolar, D., Optical properties of graphite and boron nitride. *Journal of Physics C: Solid State Physics.* **1972**, *5* (21), 3097–3100.

138. Zunger, A.; Katzir, A.; Halperin, A., Optical properties of hexagonal boron nitride. *Phys. Rev. B.* **1976**, *13* (12), 5560–5573.

139. Furthmüller, J.; Hafner, J.; Kresse, G., Ab initio calculation of the structural and electronic properties of carbon and boron nitride using ultrasoft pseudopotentials. *Phys. Rev. B.* **1994**, *50* (21), 15606–15622.

140. Blase, X.; Rubio, A.; Louie, S. G.; Cohen, M. L., Quasiparticle band structure of bulk hexagonal boron nitride and related systems. *Phys. Rev. B.* **1995**, *51* (11), 6868–6875.

141. Watanabe, K.; Taniguchi, T.; Kanda, H., Direct-bandgap properties and evidence for ultraviolet lasing of hexagonal boron nitride single crystal. *Nat. Mater.* **2004**, *3* (6), 404–409.

142. Arnaud, B.; Lebègue, S.; Rabiller, P.; Alouani, M., Huge excitonic effects in layered hexagonal boron nitride. *Phys. Rev. Lett.* **2006**, *96* (2), 026402.

143. Gao, S.-P., Crystal structures and band gap characters of h-BN polytypes predicted by the dispersion corrected DFT and GW method. *Solid State Commun.* **2012**, *152* (19), 1817–1820.

144. Paleari, F.; Galvani, T.; Amara, H.; Ducastelle, F.; Molina-Sánchez, A.; Wirtz, L., Excitons in few-layer hexagonal boron nitride: Davydov splitting and surface localization. *2D Mater.* **2018**, *5* (4), 045017.

145. Schué, L.; Berini, B.; Betz, A. C.; Plaçais, B.; Ducastelle, F.; Barjon, J.; Loiseau, A., Dimensionality effects on the luminescence properties of h-BN. *Nanoscale.* **2016**, *8* (13), 6986–6993.

146. Elias, C.; Valvin, P.; Pelini, T.; Summerfield, A.; Mellor, C. J.; Cheng, T. S.; Eaves, L.; Foxon, C. T.; Beton, P. H.; Novikov, S. V.; Gil, B.; Cassabois, G., Direct band-gap crossover in epitaxial monolayer boron nitride. *Nat. Commun.* **2019**, *10* (1), 2639.

147. Caldwell, J. D.; Aharonovich, I.; Cassabois, G.; Edgar, J. H.; Gil, B.; Basov, D. N., Photonics with hexagonal boron nitride. *Nat. Rev. Mater.* **2019**, *4* (8), 552–567.

148. Geick, R.; Perry, C. H.; Rupprecht, G., Normal modes in hexagonal boron nitride. *Phys. Rev.* **1966**, *146* (2), 543–547.

149. Ohba, N.; Miwa, K.; Nagasako, N.; Fukumoto, A., First-principles study on structural, dielectric, and dynamical properties for three BN polytypes. *Phys. Rev. B.* **2001**, *63* (11), 115207.

150. Gorbachev, R. V.; Riaz, I.; Nair, R. R.; Jalil, R.; Britnell, L.; Belle, B. D.; Hill, E. W.; Novoselov, K. S.; Watanabe, K.; Taniguchi, T.; Geim, A. K.; Blake, P., Hunting for monolayer boron nitride: Optical and Raman signatures. *Small.* **2011**, *7* (4), 465–468.

151. Harrison, H.; Lamb, J. T.; Nowlin, K. S.; Guenthner, A. J.; Ghiassi, K. B.; Kelkar, A. D.; Alston, J. R., Quantification of hexagonal boron nitride impurities in boron nitride nanotubes *via* FTIR spectroscopy. *Nanoscale Adv.* **2019**, *1* (5), 1693–1701.

152. Lee, C. H.; Jieshang, W.; Vijaya, K. K.; Huang, J. Y.; Yap, Y. K., Effective growth of boron nitride nanotubes by thermal chemical vapor deposition. *Nanotechnology.* **2008**, *19* (45), 455605.

153. Alon, O. E., Symmetry properties of single-walled boron nitride nanotubes. *Phys. Rev. B.* **2001**, *64* (15), 153408.

154. Fakrach, B.; Rahmani, A.; Chadli, H.; Sbai, K.; Bentaleb, M.; Bantignies, J. L.; Sauvajol, J. L., Infrared spectrum of single-walled boron nitride nanotubes. *Phys. Rev. B.* **2012**, *85* (11), 115437.

155. Bi, J.-Q.; Wang, W.-L.; Qi, Y.-X.; Bai, Y.-J.; Pang, L.-L.; Zhu, H.-L.; Zhao, Y.; Wang, Y., Large-scale synthesis of BN nanotubes using carbon nanotubes as template. *Materials Letters.* **2009**, *63* (15), 1299–1302.

156. Lim, S. H.; Luo, J.; Ji, W.; Lin, J., Synthesis of boron nitride nanotubes and its hydrogen uptake. *Catal. Today.* **2007**, *120* (3), 346–350.

157. Yu, J.; Chen, Y.; Cheng, B. M., Dispersion of boron nitride nanotubes in aqueous solution with the help of ionic surfactants. *Solid State Commun.* **2009**, *149* (19), 763–766.

158. Arenal, R.; Ferrari, A. C.; Reich, S.; Wirtz, L.; Mevellec, J. Y.; Lefrant, S.; Rubio, A.; Loiseau, A., Raman spectroscopy of single-walled boron nitride nanotubes. *Nano Lett.* **2006**, *6* (8), 1812–1816.

159. Wu, J.; Han, W.-Q.; Walukiewicz, W.; Ager, J. W.; Shan, W.; Haller, E. E.; Zettl, A., Raman spectroscopy and time-resolved photoluminescence of BN and BxCyNz nanotubes. *Nano Lett.* **2004**, *4* (4), 647–650.

160. Wirtz, L.; Lazzeri, M.; Mauri, F.; Rubio, A., Raman spectra of BN nanotubes: Ab initio and bond-polarizability model calculations. *Phys. Rev. B.* **2005**, *71* (24), 241402.

161. Aydin, M., Vibrational and electronic properties of single-walled and double-walled boron nitride nanotubes. *Vib. Spectrosc.* **2013**, *66*, 30–42.

162. Zhi, C.; Bando, Y.; Tang, C.; Golberg, D.; Xie, R.; Sekigushi, T., Phonon characteristics and cathodoluminінescence of boron nitride nanotubes. *Appl. Phys. Lett.* **2005**, *86* (21), 213110.

163. Chen, H.; Chen, Y.; Liu, Y.; Xu, C.-N.; Williams, J. S., Light emission and excitonic effect of boron nitride nanotubes observed by photoluminescent spectra. *Opt. Mater.* **2007**, *29* (11), 1295–1298.

164. Jaffrennou, P.; Donatini, F.; Barjon, J.; Lauret, J. S.; Maguer, A.; Attal-Tretout, B.; Ducastelle, F.; Loiseau, A., Cathodoluminescence imaging and spectroscopy on a single multiwall boron nitride nanotube. *Chem. Phys. Lett.* **2007**, *442* (4–6), 372–375.

165. Jaffrennou, P.; Barjon, J.; Schmid, T.; Museur, L.; Kanaev, A.; Lauret, J. S.; Zhi, C. Y.; Tang, C.; Bando, Y.; Golberg, D.; Attal-Tretout, B.; Ducastelle, F.; Loiseau, A., Near-band-edge recombinations in multiwalled boron nitride nanotubes: Cathodoluminescence and photoluminescence spectroscopy measurements. *Phys. Rev. B.* **2008**, *77* (23), 235422.

166. McDonald, T. J.; Engtrakul, C.; Jones, M.; Rumbles, G.; Heben, M. J., Kinetics of PL quenching during single-walled carbon nanotube rebundling and diameter-dependent surfactant interactions. *J. Phys. Chem. B.* **2006**, *110* (50), 25339–25346.

167. Huang, Y.; Bando, Y.; Tang, C.; Zhi, C.; Terao, T.; Dierre, B.; Sekiguchi, T.; Golberg, D., Thin-walled boron nitride microtubes exhibiting intense band-edge UV emission at room temperature. *Nanotechnology.* **2009**, *20* (8), 085705.

168. Huang, Y.; Lin, J.; Tang, C.; Bando, Y.; Zhi, C.; T., Z.; Dierre, B.; Sekiguchi, T.; Golberg, D., Bulk synthesis, growth mechanism and properties of highly pure ultrafine boron nitride nanotubes with diameters of sub-10 nm. *Nanotechnology.* **2011**, *22* (14), 145602.

169. Rubio, A.; Corkill, J. L.; Cohen, M. L., Theory of graphitic boron nitride nanotubes. *Phys. Rev. B.* **1994,** *49* (7), 5081–5084.

170. Blase, X.; Rubio, A.; Louie, S. G.; Cohen, M. L., Stability and band gap constancy of boron nitride nanotubes. *EPL.* **1994,** *28* (5), 335–340.

171. Chou, Y.-M.; Wang, H.-W.; Lin, Y.-J.; Chen, W.-H.; Wang, B.-C., Infinite single-walled boron-nitride nanotubes studies by LGTO-PBC-DFT method. *Diamond Relat. Mater.* **2009,** *18* (2), 351–354.

172. Yu, J.; Yu, D.; Chen, Y.; Chen, H.; Lin, M.-Y.; Cheng, B.-M.; Li, J.; Duan, W., Narrowed bandgaps and stronger excitonic effects from small boron nitride nanotubes. *Chem. Phys. Lett.* **2009,** *476* (4), 240–243.

173. Gou, G.; Pan, B.; Shi, L., The nature of radiative transitions in O-doped boron nitride nanotubes. *J. Am. Chem. Soc.* **2009,** *131* (13), 4839–4845.

174. Guerini, S.; Kar, T.; Piquini, P., Theoretical study of Si impurities in BN nanotubes. *Eur. Phys. J. B.* **2004,** *38* (3), 515–518.

175. Yan, B.; Park, C.; Ihm, J.; Zhou, G.; Duan, W.; Park, N., Electron emission originated from free-electron-like states of alkali-doped boron-nitride nanotubes. *J. Am. Chem. Soc.* **2008,** *130* (50), 17012–17015.

176. Cho, Y. J.; Kim, C. H.; Kim, H. S.; Park, J.; Choi, H. C.; Shin, H.-J.; Gao, G.; Kang, H. S., Electronic structure of Si-doped BN nanotubes using X-ray photoelectron spectroscopy and first-principles calculation. *Chem. Mater.* **2009,** *21* (1), 136–143.

177. Chen, C.-W.; Lee, M.-H.; Clark, S. J., Band gap modification of single-walled carbon nanotube and boron nitride nanotube under a transverse electric field. *Nanotechnology.* **2004,** *15* (12), 1837–1843.

178. Lauret, J. S.; Arenal, R.; Ducastelle, F.; Loiseau, A.; Cau, M.; Attal-Tretout, B.; Rosencher, E.; Goux-Capes, L., Optical transitions in single-walled boron nitride nanotubes. *Phys. Rev. Lett.* **2005,** *94* (3), 037405.

179. Arenal, R.; Stéphan, O.; Kociak, M.; Taverna, D.; Loiseau, A.; Colliex, C., Electron energy loss spectroscopy measurement of the optical gaps on individual boron nitride single-walled and multiwalled nanotubes. *Phys. Rev. Lett.* **2005,** *95* (12), 127601.

180. Wirtz, L.; Marini, A.; Rubio, A., Excitons in boron nitride nanotubes: Dimensionality effects. *Phys. Rev. Lett.* **2006,** *96* (12), 126104.

181. Park, C.-H.; Spataru, C. D.; Louie, S. G., Excitons and many-electron effects in the optical response of single-walled boron nitride nanotubes. *Phys. Rev. Lett.* **2006,** *96* (12), 126105.

182. Wang, L.; Xu, X.; Zhang, L.; Qiao, R.; Wu, M.; Wang, Z.; Zhang, S.; Liang, J.; Zhang, Z.; Zhang, Z.; Chen, W.; Xie, X.; Zong, J.; Shan, Y.; Guo, Y.; Willinger, M.; Wu, H.; Li, Q.; Wang, W.; Gao, P.; Wu, S.; Zhang, Y.; Jiang, Y.; Yu, D.; Wang, E.; Bai, X.; Wang, Z.-J.; Ding, F.; Liu, K., Epitaxial growth of a 100-square-centimetre single-crystal hexagonal boron nitride monolayer on copper. *Nature.* **2019**, *570* (7759), 91–95.

183. Jeong, H.; Kim, D. Y.; Kim, J.; Moon, S.; Han, N.; Lee, S. H.; Okello, O. F. N.; Song, K.; Choi, S.-Y.; Kim, J. K., Wafer-scale and selective-area growth of high-quality hexagonal boron nitride on Ni(111) by metal-organic chemical vapor deposition. *Sci. Rep.* **2019**, *9* (1), 5736.

184. Zhang, X. W., Doping and electrical properties of cubic boron nitride thin films: A critical review. *Thin Solid Films.* **2013**, *544*, 2–12.

185. Yankowitz, M.; Ma, Q.; Jarillo-Herrero, P.; LeRoy, B. J., van der Waals heterostructures combining graphene and hexagonal boron nitride. *Nat. Rev. Phys.* **2019**, *1* (2), 112–125.

186. Kim, K. K.; Lee, H. S.; Lee, Y. H., Synthesis of hexagonal boron nitride heterostructures for 2D van der Waals electronics. *Chem. Soc. Rev.* **2018**, *47* (16), 6342–6369.

187. Wang, J.; Ma, F.; Sun, M., Graphene, hexagonal boron nitride, and their heterostructures: Properties and applications. *RSC Adv.* **2017**, *7* (27), 16801–16822.

188. Lee, C.; Bhandari, S.; Tiwari, B.; Yapici, N.; Zhang, D.; Yap, Y. K., Boron nitride nanotubes: Recent advances in their synthesis, functionalization, and applications. *Molecules.* **2016**, *21* (7), 922.

189. Nose, K.; Oba, H.; Yoshida, T., Electric conductivity of boron nitride thin films enhanced by *in situ* doping of zinc. *Appl. Phys. Lett.* **2006**, *89* (11), 112124.

190. He, B.; Zhang, W. J.; Zou, Y. S.; Chong, Y. M.; Ye, Q.; Ji, A. L.; Yang, Y.; Bello, I.; Lee, S. T.; Chen, G. H., Electrical properties of Be-implanted polycrystalline cubic boron nitride films. *Appl. Phys. Lett.* **2008**, *92* (10), 102108.

191. Taniguchi, T.; Teraji, T.; Koizumi, S.; Watanabe, K.; Yamaoka, S., Appearance of n-type semiconducting properties of cBN single crystals grown at high pressure. *Jpn. J. Appl. Phys.* **2002**, *41* (Part 2, No. 2A), L109–L111.

192. Wang, C.-X.; Yang, G.-W.; Zhang, T.-C.; Liu, H.-W.; Han, Y.-H.; Luo, J.-F.; Gao, C.-X.; Zou, G.-T., High-quality heterojunction between p-type diamond single-crystal film and n-type cubic boron nitride bulk single crystal. *Appl. Phys. Lett.* **2003**, *83* (23), 4854–4856.

193. Sun, F.; Hao, Z.; Liu, G.; Wu, C.; Lu, S.; Huang, S.; Liu, C.; Hong, Q.; Chen, X.; Cai, D.; Kang, J., p-Type conductivity of hexagonal boron nitride as a dielectrically tunable monolayer: Modulation doping with magnesium. *Nanoscale.* **2018**, *10* (9), 4361–4369.

194. Lu, M.; Bousetta, A.; Bensaoula, A.; Waters, K.; Schultz, J. A., Electrical properties of boron nitride thin films grown by neutralized nitrogen ion assisted vapor deposition. *App. Phys. Lett.* **1996**, *68* (5), 622–624.

195. Han, W.; Bando, Y.; Kurashima, K.; Sato, T., Synthesis of boron nitride nanotubes from carbon nanotubes by a substitution reaction. *Appl. Phys. Lett.* **1998**, *73* (21), 3085–3087.

196. Wentorf, R. H., Preparation of semiconducting cubic boron nitride. *J. Chem. Phys.* **1962**, *36* (8), 1990–1991.

197. Mishima, O.; Era, K.; Tanaka, J.; Yamaoka, S., Ultraviolet light-emitting diode of a cubic boron nitride pn junction made at high pressure. *Appl. Phys. Lett.* **1988**, *53* (11), 962–964.

198. Litvinov, D.; Taylor, C. A.; Clarke, R., Semiconducting cubic boron nitride. *Diamond Relat. Mater.* **1998**, *7* (2), 360–364.

199. Nose, K.; Tachibana, K.; Yoshida, T., Rectification properties of layered boron nitride films on silicon. *Appl. Phys. Lett.* **2003**, *83* (5), 943–945.

200. Gubanov, V. A.; Lu, Z. W.; Klein, B. M.; Fong, C. Y., Electronic structure of defects and impurities in III-V nitrides: Vacancies in cubic boron nitride. *Phys. Rev. B.* **1996**, *53* (8), 4377–4385.

201. Orellana, W.; Chacham, H., Atomic geometry and energetics of vacancies and antisites in cubic boron nitride. *Appl. Phys. Lett.* **1999**, *74* (20), 2984–2986.

202. Ying, J.; Zhang, X. W.; Yin, Z. G.; Tan, H. R.; Zhang, S. G.; Fan, Y. M., Electrical transport properties of the Si-doped cubic boron nitride thin films prepared by *in situ* cosputtering. *J. Appl. Phys.* **2011**, *109* (2), 023716.

203. Bello, I.; Chong, Y. M.; Ye, Q.; Yang, Y.; He, B.; Kutsay, O.; Wang, H. E.; Yan, C.; Jha, S. K.; Zapien, J. A.; Zhang, W. J., Materials with extreme properties: Their structuring and applications. *Vacuum.* **2012**, *86* (6), 575–585.

204. Yin, H.; Pongrac, I.; Ziemann, P., Electronic transport in heavily Si doped cubic boron nitride films epitaxially grown on diamond (001). *J. Appl. Phys.* **2008**, *104* (2), 023703.

205. Ying, J.; Zhang, X. W.; Fan, Y. M.; Tan, H. R.; Yin, Z. G., Effects of silicon incorporation on composition, structure and electric conductivity of cubic boron nitride thin films. *Diamond Relat. Mater.* **2010**, *19* (11), 1371–1376.

206. Ronning, C.; Banks, A. D.; McCarson, B. L.; Schlesser, R.; Sitar, Z.; Davis, R. F.; Ward, B. L.; Nemanich, R. J., Structural and electronic properties of boron nitride thin films containing silicon. *J. Appl. Phys.* **1998**, *84* (9), 5046–5051.

207. Taniguchi, T.; Koizumi, S.; Watanabe, K.; Sakaguchi, I.; Sekiguchi, T.; Yamaoka, S., High pressure synthesis of UV-light emitting cubic boron nitride single crystals. *Diamond Relat. Mater.* **2003**, *12* (3), 1098–1102.

208. Kojima, K.; Nose, K.; Kambara, M.; Yoshida, T., Effects of magnesium doping on growth and electric conductivity of nanocrystalline cubic boron nitride thin films. *J. Phys. D: Appl. Phys.* **2009**, *42* (5), 055304.

209. Li, J.; Gui, G.; Zhong, J., Tunable bandgap structures of two-dimensional boron nitride. *J. Appl. Phys.* **2008**, *104* (9), 094311.

210. Li, L. H.; Santos, E. J. G.; Xing, T.; Cappelluti, E.; Roldán, R.; Chen, Y.; Watanabe, K.; Taniguchi, T., Dielectric screening in atomically thin boron nitride nanosheets. *Nano Lett.* **2015**, *15* (1), 218–223.

211. Carpenter, L. G.; Kirby, P. J., The electrical resistivity of boron nitride over the temperature range 700 degrees C to 1400 degrees C. *J. Phys. D: Appl. Phys.* **1982**, *15* (7), 1143–1151.

212. Uddin, M. R.; Majety, S.; Li, J.; Lin, J. Y.; Jiang, H. X., Layer-structured hexagonal (BN)C semiconductor alloys with tunable optical and electrical properties. *J. Appl. Phys.* **2014**, *115* (9), 093509.

213. He, B.; Zhang, W. J.; Yao, Z. Q.; Chong, Y. M.; Yang, Y.; Ye, Q.; Pan, X. J.; Zapien, J. A.; Bello, I.; Lee, S. T.; Gerhards, I.; Zutz, H.; Hofsäss, H., *p*-Type conduction in beryllium-implanted hexagonal boron nitride films. *Appl. Phys. Lett.* **2009**, *95* (25), 252106.

214. Lu, M.; Bousetta, A.; Bensaoula, A.; Waters, K.; Schultz, J. A., Electrical properties of boron nitride thin films grown by neutralized nitrogen ion assisted vapor deposition. *Applied Physics Letters.* **1996**, *68* (5), 622–624.

215. Xue, Y.; Liu, Q.; He, G.; Xu, K.; Jiang, L.; Hu, X.; Hu, J., Excellent electrical conductivity of the exfoliated and fluorinated hexagonal boron nitride nanosheets. *Nanoscale Res. Lett.* **2013**, *8* (1), 49.

216. Huang, J. Y.; Yasuda, H.; Mori, H., HRTEM and EELS studies on the amorphization of hexagonal boron nitride induced by Ball Milling. *J. Am. Ceram. Soc.* **2000**, *83* (2), 403–409.

217. Arenal, R.; Kociak, M.; Zaluzec, N. J., High-angular-resolution electron energy loss spectroscopy of hexagonal boron nitride. *Appl. Phys. Lett.* **2007**, *90* (20), 204105.

218. Giovannetti, G.; Khomyakov, P. A.; Brocks, G.; Kelly, P. J.; van den Brink, J., Substrate-induced band gap in graphene on hexagonal boron nitride: Ab initio density functional calculations. *Phys. Rev. B.* **2007**, *76* (7), 073103.

219. Dean, C. R.; Young, A. F.; Meric, I.; Lee, C.; Wang, L.; Sorgenfrei, S.; Watanabe, K.; Taniguchi, T.; Kim, P.; Shepard, K. L.; Hone, J., Boron nitride substrates for high-quality graphene electronics. *Nature Nanotechnology.* **2010**, *5* (10), 722–726.

220. Kim, K. K.; Hsu, A.; Jia, X.; Kim, S. M.; Shi, Y.; Dresselhaus, M.; Palacios, T.; Kong, J., Synthesis and characterization of hexagonal boron nitride film as a dielectric layer for graphene devices. *ACS Nano.* **2012**, *6* (10), 8583–8590.

221. Kumar, P.; Chauhan, Y. S.; Agarwal, A.; Bhowmick, S., Thickness and Stacking dependent polarizability and dielectric constant of graphene–hexagonal boron nitride composite stacks. *J. Phys. Chem. C.* **2016**, *120* (31), 17620–17626.

222. Zomer, P. J.; Dash, S. P.; Tombros, N.; van Wees, B. J., A transfer technique for high mobility graphene devices on commercially available hexagonal boron nitride. *Appl. Phys. Lett.* **2011**, *99* (23), 232104.

223. Bresnehan, M. S.; Hollander, M. J.; Wetherington, M.; LaBella, M.; Trumbull, K. A.; Cavalero, R.; Snyder, D. W.; Robinson, J. A., Integration of hexagonal boron nitride with quasi-freestanding epitaxial graphene: Toward wafer-scale, high-performance devices. *ACS Nano.* **2012**, *6* (6), 5234–5241.

224. Aggoune, W.; Cocchi, C.; Nabok, D.; Rezouali, K.; Belkhir, M. A.; Draxl, C., Structural, electronic, and optical properties of periodic graphene/h-BN van der Waals heterostructures. *Physical Review Materials.* **2020**, *4* (8), 084001.

225. Zhang, C.; Zhao, S.; Jin, C.; Koh, A. L.; Zhou, Y.; Xu, W.; Li, Q.; Xiong, Q.; Peng, H.; Liu, Z., Direct growth of large-area graphene and boron nitride heterostructures by a co-segregation method. *Nat. Commun.* **2015**, *6*, 6519.

226. Withers, F.; Bointon, T. H.; Hudson, D. C.; Craciun, M. F.; Russo, S., Electron transport of WS2 transistors in a hexagonal boron nitride dielectric environment. *Sci. Rep.* **2014**, *4* (1), 4967.

227. Chan, M. Y.; Komatsu, K.; Li, S.-L.; Xu, Y.; Darmawan, P.; Kuramochi, H.; Nakaharai, S.; Aparecido-Ferreira, A.; Watanabe, K.; Taniguchi, T.; Tsukagoshi, K., Suppression of thermally activated carrier transport in atomically thin MoS2 on crystalline hexagonal boron nitride substrates. *Nanoscale.* **2013**, *5* (20), 9572–9576.

228. Das, S.; Gulotty, R.; Sumant, A. V.; Roelofs, A., All two-dimensional, flexible, transparent, and thinnest thin film transistor. *Nano Lett.* **2014**, *14* (5), 2861–2866.

229. Laturia, A.; Van de Put, M. L.; Vandenberghe, W. G., Dielectric properties of hexagonal boron nitride and transition metal dichalcogenides: From monolayer to bulk. *npj 2D Mater. Appl.* **2018**, *2* (1), 6.

230. Andersen, K.; Latini, S.; Thygesen, K. S., Dielectric genome of van der Waals heterostructures. *Nano Lett.* **2015**, *15* (7), 4616–4621.

231. Latini, S.; Olsen, T.; Thygesen, K. S., Excitons in van der Waals heterostructures: The important role of dielectric screening. *Phys. Rev. B.* **2015**, *92* (24), 245123.

232. Gjerding, M. N.; Cavalcante, L. S. R.; Chaves, A.; Thygesen, K. S., Efficient ab initio modeling of dielectric screening in 2D van der Waals materials: Including phonons, substrates, and doping. *J. Phys. Chem. C.* **2020**, *124* (21), 11609–11616.

233. Raja, A.; Waldecker, L.; Zipfel, J.; Cho, Y.; Brem, S.; Ziegler, J. D.; Kulig, M.; Taniguchi, T.; Watanabe, K.; Malic, E.; Heinz, T. F.; Berkelbach, T. C.; Chernikov, A., Dielectric disorder in two-dimensional materials. *Nat. Nanotech.* **2019**, *14* (9), 832–837.

234. Steinleitner, P.; Merkl, P.; Graf, A.; Nagler, P.; Watanabe, K.; Taniguchi, T.; Zipfel, J.; Schüller, C.; Korn, T.; Chernikov, A.; Brem, S.; Selig, M.; Berghäuser, G.; Malic, E.; Huber, R., Dielectric engineering of electronic correlations in a van der Waals heterostructure. *Nano Lett.* **2018**, *18* (2), 1402–1409.

235. Winther, K. T.; Thygesen, K. S., Band structure engineering in van der Waals heterostructures *via* dielectric screening: The GΔW method. *2D Mater.* **2017**, *4* (2), 025059.

236. Crespi, V. H.; Cohen, M. L.; Rubio, A., *In situ* band gap engineering of carbon nanotubes. *Phys. Rev. Lett.* **1997**, *79* (11), 2093–2096.

237. Zhuang, C.; Xu, H.; Li, L.; Liu, Y.; Ban, C.; Liu, X., Systematic investigation of the ball milling–annealing growth and electrical properties of boron nitride nanotubes. *RSC Adv.* **2016**, *6* (114), 113415–113423.

238. Cumings, J.; Zettl, A., Field emission and current-voltage properties of boron nitride nanotubes. *Solid State Commun.* **2004**, *129* (10), 661–664.

239. Miyamoto, Y.; Rubio, A.; Cohen, M. L.; Louie, S. G., Chiral tubules of hexagonal BC2N. *Phys. Rev. B.* **1994**, *50* (7), 4976–4979.

240. Watanabe, M. O.; Itoh, S.; Sasaki, T.; Mizushima, K., Visible-light-emitting layered BC_2N Semiconductor. *Phys. Rev. Lett.* **1996**, *77* (1), 187–189.

241. Weng, Q.; Kvashnin, D. G.; Wang, X.; Cretu, O.; Yang, Y.; Zhou, M.; Zhang, C.; Tang, D.-M.; Sorokin, P. B.; Bando, Y.; Golberg, D., Tuning of the optical, electronic, and magnetic properties of boron nitride nanosheets with oxygen doping and functionalization. *Adv. Mater.* **2017**, *29* (28), 1700695.

242. Blase, X.; Charlier, J. C.; De Vita, A.; Car, R., Theory of composite BxCyNz nanotube heterojunctions. *Appl. Phys. Lett.* **1997**, *70* (2), 197–199.

243. Suenaga, K.; Colliex, C.; Demoncy, N.; Loiseau, A.; Pascard, H.; Willaime, F., Synthesis of nanoparticles and nanotubes with well-separated layers of boron nitride and carbon. *Science.* **1997**, *278* (5338), 653.

244. Mickelson, W.; Aloni, S.; Han, W.-Q.; Cumings, J.; Zettl, A., Packing C_{60} in boron nitride nanotubes. *Science.* **2003**, *300* (5618), 467.

245. Yoshioka, T.; Suzuura, H.; Ando, T., Electronic states of BCN alloy nanotubes in a simple tight binding model. *J. Phys. Soc. Jpn.* **2003**, *72* (10), 2656–2664.

246. Wang, W.; Bando, Y.; Zhi, C.; Fu, W.; Wang, E.; Golberg, D., Aqueous noncovalent functionalization and controlled near-surface carbon doping of multiwalled boron nitride nanotubes. *J. Am. Chem. Soc.* **2008**, *130* (26), 8144–8145.

247. Tang, C.; Bando, Y.; Huang, Y.; Yue, S.; Gu, C.; Xu, F.; Golberg, D., Fluorination and electrical conductivity of BN nanotubes. *J. Am. Chem. Soc.* **2005**, *127* (18), 6552–6553.

248. Lai, L.; Song, W.; Lu, J.; Gao, Z.; Nagase, S.; Ni, M.; Mei, W. N.; Liu, J.; Yu, D.; Ye, H., Structural and electronic properties of fluorinated boron nitride nanotubes. *J. Phys. Chem. B.* **2006**, *110* (29), 14092–14097.

249. Xiang, H. J.; Yang, J.; Hou, J. G.; Zhu, Q., Are fluorinated boron nitride nanotubes n-type semiconductors? *Appl. Phys. Lett.* **2005**, *87* (24), 243113.

250. Zhou, Z.; Zhao, J.; Chen, Z.; Schleyer, P. v. R., Atomic and electronic structures of fluorinated BN nanotubes: Computational study. *J. Phys. Chem. B.* **2006**, *110* (51), 25678–25685.

251. Du, M.; Li, X.; Wang, A.; Wu, Y.; Hao, X.; Zhao, M., One-step exfoliation and fluorination of boron nitride nanosheets and a study of their magnetic properties. *Angew. Chem. Int. Ed.* **2014**, *53* (14), 3645–3649.

252. Hu, C.; Michaud-Rioux, V.; Yao, W.; Guo, H., Theoretical design of topological heteronanotubes. *Nano Lett.* **2019**, *19* (6), 4146–4150.

253. Sengupta, A., Atomistic study of electrostatics and carrier transport properties of CNT@MS2 (M = Mo, W) and CNT@BN core–shell nanotubes. *J. Mater. Sci.* **2017**, *52* (13), 8119–8131.

254. Suenaga, K.; Colliex, C.; Demoncy, N.; Loiseau, A.; Pascard, H.; Willaime, F., Synthesis of nanoparticles and nanotubes with well-separated layers of boron nitride and carbon. *Science.* **1997**, *278* (5338), 653–655.

255. Chen, L.; Ye, H.; Gogotsi, Y., Synthesis of boron nitride coating on carbon nanotubes. *J. Am. Ceram. Soc.* **2004**, *87* (1), 147–151.

256. Xiang, R.; Inoue, T.; Zheng, Y.; Kumamoto, A.; Qian, Y.; Sato, Y.; Liu, M.; Tang, D.; Gokhale, D.; Guo, J.; Hisama, K.; Yotsumoto, S.; Ogamoto, T.; Arai, H.; Kobayashi, Y.; Zhang, H.; Hou, B.; Anisimov, A.; Maruyama, M.; Miyata, Y.; Okada, S.; Chiashi, S.; Li, Y.; Kong, J.; Kauppinen, E. I.; Ikuhara, Y.; Suenaga, K.; Maruyama, S., One-dimensional van der Waals heterostructures. *Science.* **2020**, *367* (6477), 537–542.

257. Lan, H.-P.; Ye, L.-H.; Zhang, S.; Peng, L.-M., Transverse dielectric properties of boron nitride nanotubes by ab initio electric field calculations. *Appl. Phys. Lett.* **2009**, *94* (18), 183110.

258. Brożek, T.; Szmidt, J.; Jakubowski, A.; Olszyna, A., Electrical behaviour and breakdown in plasma deposited cubic BN layers. *Diamond Relat. Mater.* **1994**, *3* (4), 720–724.

259. Hattori, Y.; Taniguchi, T.; Watanabe, K.; Nagashio, K., Layer-by-layer dielectric breakdown of hexagonal boron nitride. *ACS Nano.* **2015**, *9* (1), 916–921.

Chapter 3

Fabrication of BN Nanomaterials

3.1 Introduction

The synthesis of BN nanomaterials is challenging in practice. An extra chemical element comes with an additional layer of complexity compared to the synthesis of structurally analogous carbon nanomaterials. However, the practical approaches remain similar. Drawing from the experience of synthesizing carbon nanomaterials, progress was made relatively quickly regarding the synthesis of BN nanomaterials. However, with a great divide in the attention received by BN nanomaterial growth compared to that of carbon, a thorough understanding of the growth process is still lacking, despite recent breakthroughs. This chapter discusses the various synthesis methods to obtain BN nanomaterials and the current understanding of BN nanomaterial growth mechanisms.

3.2 Fabrication of c-BN

First synthesized through the application of extreme temperature and pressure to h-BN by Wentorf,[1, 2] early efforts at synthesizing c-BN employed high-energy ions bombarding h-BN or a B substrate in an atmosphere of Ar or N_2 ions. Many early methods of c-BN synthesis, including pulsed laser deposition (PLD) and sputtering,

Boron Nitride Nanomaterials: Properties, Fabrication, and Applications
Ben McLean and Alister J. Page

ISBN 978-981-4968-23-2 (Hardcover), 978-1-003-31448-6 (eBook)
www.jennystanford.com

utilize ion bombardment. Ion bombardment influences growth in myriad ways.[3] Firstly, ions such as Ar^+ become implanted in the near-surface region of the h-BN target, which results in the generation of stress and densification, increasing the number of sp^3-hybridized bonds. High-energy ions are ejected from the near-surface region inducing structural relaxation, though this may lower the c-BN content of the growing film. These methods are difficult to control, and consequently, a more controlled approach involving plasma-assisted radio frequency (RF) CVD eventually arose for the synthesis of c-BN. Several other methods have also been prominent, including CVD, arc discharge, sputtering, and evaporation, among others, as we detail below.[4–6]

3.2.1 Conversion of h-BN to c-BN

The first direct conversion of h-BN to c-BN was reported by Wentorf and co-workers,[1, 2] and their process was subsequently improved upon by Wakatsuki *et al.*[7] using a lower pressure. The latter authors also showed that the h-BN precursor particle size and crystallinity determine the conditions required to produce c-BN during the destructive-reconstructive diffusion process. In particular, smaller particle sizes and low crystallinity decrease the activation barrier.[8] This result was supported by Corrigan and Bundy[9] and Sirota[10] at higher temperatures. The a-BN[11] and t-BN[12] have also been used as starting materials, but require lower and higher temperatures for the conversion to c-BN, respectively, as compared to h-BN.

Commercialization of this technique would require milder conditions than those used by Wentorf and Wakatsuki *et al.* as high temperature and high-pressure synthesis (1,500 °C and 45,000 atm) for mass production would be expensive. Toward this end, chemical precursors were added to h-BN to form a eutectic phase that could be converted to c-BN at less extreme temperatures and pressures. These precursors include nitrides of alkali and alkali earth metals, fluoronitrides, silicon, and aluminum nitrides as well as ammonium borate (see Vel *et al.*[4] and references therein). The inclusion of these precursors facilitated the nucleation of c-BN following the decomposition of h-BN. Pressures and temperatures required were reduced to around 4 GPa and 1,500 K. Nevertheless, the investigation of h-BN/c-BN conversion synthesis remained an active avenue of research, ultimately producing c-BN-based dense BN polymorphs at a range of temperatures (1,500–2,500 K) and pressures (5–20 GPa) and produced the first non-carbon material to approach the hardness of diamond (85 GPa).[13] Similarly, a nanotwinned c-BN material produced from onion-like t-BN at 15 GPa and 2,073 K exhibited the hardness exceeding that of synthetic diamond (100 GPa).[14]

Conversion of h-BN to c-BN can also be achieved *via* an ion-assisted PLD method. Initially, a CO_2 laser ablates h-BN, with a flux of N_2^+ and Ar^+ ions helping to deposit the BN precursors onto a substrate, often Si.[15–19] More recently, a laser has been employed to melt an h-BN precursor close to its triple point such that pure c-BN phase nucleates and epitaxially grows on a sapphire substrate.[20, 21] This method allows for the c-BN content to be accurately controlled, hence tuning the optical band gap for optoelectronic applications.[22] This method, as with any technique using ion sources, is limited, however, by the difficulty of scale-up.

3.2.2 Ion Bombardment

c-BN films can be synthesized *via* bombardment of a boron substrate with electrons and the ionization of N_2 using a Penning ionization gauge (PIG) on a tantalum substrate.[23] The process itself is based on one developed for the formation of nitride films, and a schematic of a typical ion-bombardment apparatus is depicted in Fig. 3.1.[24] c-BN film formation *via* this technique occurs within a specific ion flux and ion energy ranges, which can be controlled *via* the applied substrate bias, as shown in Fig. 3.2a. The mass ratio of ions used (Ar^+ to N_2^+) can also be used to tune the c-BN content of the synthesized film.[25] Distinct thresholds for these experimental parameters controlling the BN phase (i.e., c-BN, h-BN, t-BN, and a-BN) produced were observed by numerous groups.[26–32] For instance, the effect of N/Ar ion energy and current density was investigated by Tanabe *et al.*[31] (Fig. 3.2cb), while Reinke *et al.*[33] considered substrate temperature (Fig. 3.2b) and Mirkarimi *et al.*[15] examined the role of the energy deposited in the film by the ions for different ion energies (Fig. 3.2d). This broad parameter space was then narrowed through several experimental investigations by Reinke *et al.*[33–35] Key observations that the substrate bias could be decreased upon c-BN nucleation, and that the ion energy required for nucleation was lower than that for growth, were made by McKenzie *et al.*[36] and Hahn *et al.*[37] The use of high-energy ions to form c-BN was commonplace throughout the 1990s and was studied extensively during this time (see Mirkarimi *et al.*[5] and references therein). However, due to the harsh reaction conditions, lack of experimental control, and issues with c-BN film adhesion and thickness, alternative methods were soon explored, as discussed below.

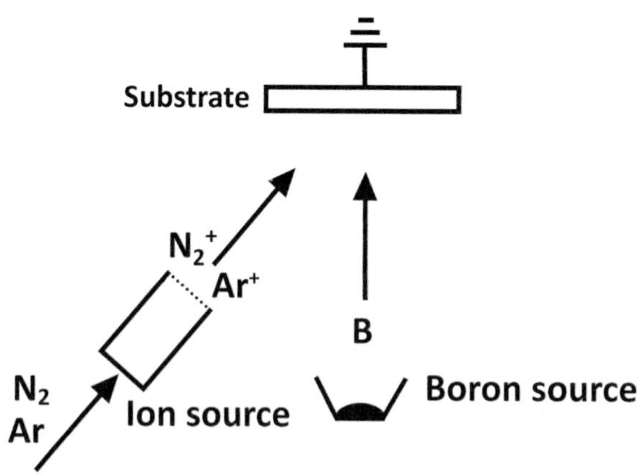

Figure 3.1 Simplified schematic illustrating ion bombardment apparatus for synthesizing c-BN films. A solid boron precursor is evaporated toward the growth substrate with an applied potential, where it is combined with nitrogen provided by the ion source.[5] Reprinted from Mirkarimi, P. B. *et al.* Review of advances in cubic boron nitride film synthesis. *Mater. Sci. Eng. R* 1997, 21 (2), 47–100. Copyright 1997, with permission from Elsevier.

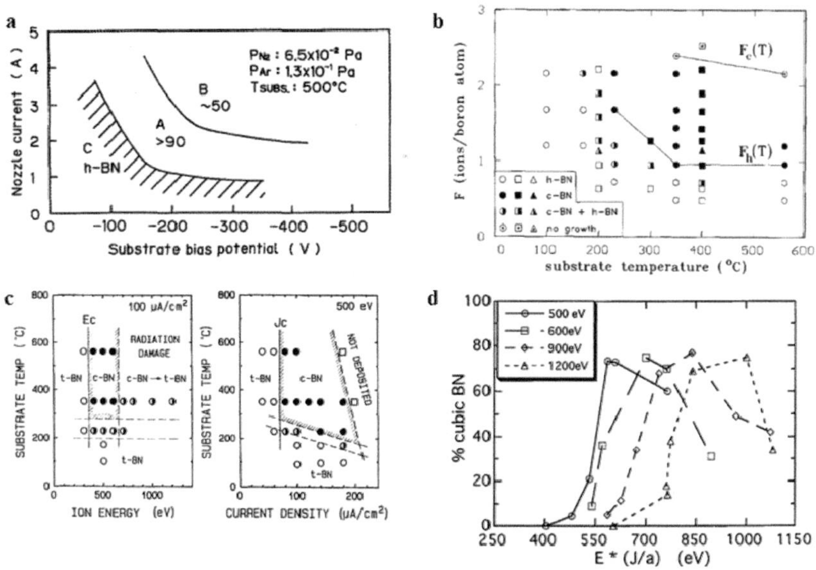

Figure 3.2 Dependence of deposited BN film phase on experimental parameters using ion bombardment methods. (a) The percentage of c-BN film content as a function of applied substrate bias potential controlling the bombardment of the B substrate with nitrogen from a hollow cathode discharge.[25] Reprinted from Inagawa, K. *et al.* Preparation of cubic boron nitride film by activated reactive evaporation with a gas activation nozzle. *J. Vac. Sci. Technol. A* 1987, 5 (4), 2696–2700, with the permission of AIP Publishing. (b) Deposited phase as a function of substrate temperature by (lines) Reinke *et al.*[33] compared to (circles) Tanabe *et al.*,[31] (squares) Kester *et al.*,[27] and (triangles) Djouadi *et al.*[38] Reprinted from Reinke, S. *et al.* Mechanisms in ion-induced c-BN growth. *Diamond Relat. Mater.* 1994, 3 (4), 341–345. Copyright 1994, with permission from Elsevier. (c) BN phases are presented as a function of N/Ar ion energy and Si substrate temperature.[31] Reprinted from Tanabe *et al.* Substrate temperature influence of c-BN thin film formation by IBED. *Diamond Relat. Mater.* 1992, 1 (2), 151–156. Copyright 1992, with permission from Elsevier. (d) c-BN content produced from N/Ar energies between 500 eV and 1,200 eV as a function of the energy deposited by the ions bombarding the film per atom.[15] Reprinted by permission from Springer Nature Customer Service Center GmbH: Springer Nature. Mirkarimi, P. B. *et al.* On the role of ions in the formation of cubic boron nitride films by ion-assisted deposition. *J. Mater. Res.* 1994, 9 (11), 2925–2938. Copyright 1994.

3.2.3 Sputtering

RF magnetron sputtering was amongst the first alternatives to high-energy ion-assisted deposition of c-BN films. Their development followed experiments by Mieno and Yoshida that produced pure c-BN film by sputtering an h-BN target in a pure Ar discharge.[39] Sputtering approaches were pursued extensively

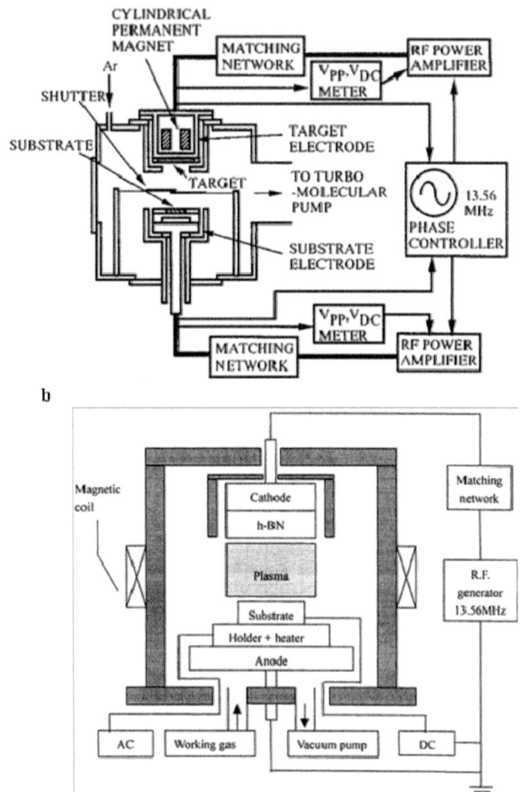

Figure 3.3 (a) RF bias sputtering utilizing an unbalanced magnetron configuration.[42] Reprinted from Tsuda, O. *et al.* Preparation of cubic boron nitride films by radio frequency bias sputtering. *J. Vac. Sci. Technol. A* 1995, 13 (6), 2843–2847, with the permission of AIP Publishing. (b) RF bias sputtering using a magnetic coil to confine the movement of charged particles to the plasma region between the h-BN target and Si substrate.[49] Reprinted from Deng, J. *et al.* The growth of cubic boron nitride films by RF reactive sputter. *Thin Solid Films* 2000, 368 (2), 312–314. Copyright 2000, with permission from Elsevier.

for much of the 1990s and early 2000s, and resulted in films with high c-BN content (e.g., upwards of 85%).[40–54] This approach would often be paired with ion bombardment, to preferentially remove h-BN over c-BN during synthesis, and also to provide an additional source of N, if needed. Apparatuses used to synthesize c-BN are shown schematically in Fig. 3.3. Figure 3.3a shows a direct current (DC) sputtering setup that employed B_4C targets in an N_2 atmosphere; while able to produce c-BN, carbon impurities persisted in the final c-BN film due to its presence in the B_4C precursor.[55–57] Figure 3.3b shows RF sputtering systems that deposit c-BN onto Si from an h-BN target using two different magnetic configurations to direct the magnetic force to the substrate. The addition of H_2 reduces the compressive stress of the c-BN film.[58, 59] The required substrate bias for DC sputtering, as opposed to RF sputtering, was three times higher[60] and often produced films of lower c-BN content.[41] The principal disadvantage of these sputtering techniques is the lack of experimental control over the final synthesized product. This is particularly the case for the ion-assisted method, as the flux of ions and sputtered material is difficult to control and measure independently.

3.2.4 Plasma-Assisted Chemical Vapor Deposition

With c-BN's structural similarities to diamond, the use of plasma-assisted CVD for c-BN synthesis saw widespread research interest following its success as a method for synthesizing diamond films.[61–64] Indeed, it was successful in an early study synthesizing a BN/diamond layered structure.[65] Chayahara *et al.*,[66–68] Yokoyama *et al.*,[26, 69] Okamoto *et al.*,[70] and Osaka *et al.*[71] utilized PACVD to grow c-BN films on Si from a diborane (B_2H_6) precursor in an atmosphere of N_2 and Ar. An RF discharge was applied at low pressures under a magnetic field, to confine the plasma for negatively self-biased substrate electrodes. Much like the sputtering and ion-bombardment synthetic approaches discussed above, ion bombardment of the growth surface during PACVD was found to be the key for the formation of a cubic BN phase. The growth of c-BN occurred only in a narrow range of self-bias (−200 to −300 V) in these studies,[66] as well as others,[72–80] and a distinction between the conditions required for nucleation and growth has also been reported.[81] Dworschak *et al.*[72, 73] saw an increase in the c-BN:h-BN ratio by maintaining the B:N ratio in the growth material below 1.1 and the Ar concentration higher than that of the H in the plasma (Fig. 3.4a). Results suggested that H incorporated into the growing film and limited the growth of the c-BN phase. Experiments employing ion-assisted evaporation[27] and first-principle calculations[15] instead indicate that this may simply be due to the lack of sufficient momentum transfer to the substrate to induce stress and facilitate c-BN growth, rather than any chemical effect induced by the presence of hydrogen. Saitoh *et al.*[82–84] and Ichinose *et al.*[85] reported that the use of solid precursors during thermally assisted microwave PACVD (Fig. 3.4b), such as ammonia borane (NH_3-BH_3), H_3BO_3, and $NaBH_4$ resulted in an increase in the c-BN content due to h-BN etching by H. Similar observations were made by Loeffler[86] and Loh.[87] Ichinose *et al.* reported significant progress in the design of their CVD apparatus; the use of a W filament over a substrate bias reportedly decomposes the solid precursors and allows for BNH radical species to deposit on the Si or diamond substrate, before nucleation of pure c-BN takes place. The use of a Ni catalyst substrate[88, 89] and a stainless steel substrate containing Ni[90] in a similar system also saw high-purity c-BN formed.

Fluorine subsequently emerged as an effective chemical etchant for h-BN and a-BN during PACVD synthesis of c-BN.[91] Kalss *et al.*[92] showed that h-BN was etched six times faster than c-BN using a BF_3 precursor in place of BH_3 under equivalent synthesis conditions. Using the same chemical precursors, Zhang *et al.*[93, 94] and Matsumoto *et al.*[95] observed 90% c-BN content on 25 μm thick films grown in Si (001), proposing that BF_3-derived F acted as the principal etchant, while H stabilized the c-BN surface *via* chemical passivation. The use of a DC jet to assist growth more effectively produced films with high c-BN content, compared to microwaves.[95, 96] The first clear crystal facets of c-BN were observed using this DC jet PACVD method by both Zhang *et al.*[97] and Matsumoto *et al.*[98] Here the substrate bias was increased from −85 V to −70 V following the nucleation of c-BN for optimal film growth. Continued collaborative efforts from Zhang and Matsumoto[6, 99–101] saw high-quality c-BN (>80%) films produced on Si (001) using PACVD. Electron cyclotron

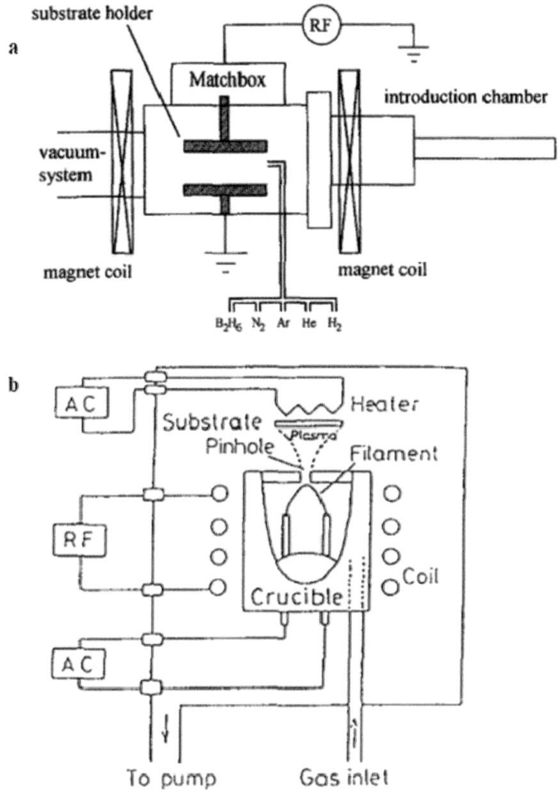

Figure 3.4 (a) Horizontal PACVD reactor split into two chambers: introduction and deposition, with a potential bias able to be applied to the substrate.[72] Reprinted from Dworschak, W. *et al.* Growth of cubic boron nitride coatings in a magnetic field enhanced r.f. glow discharge. *Diamond Relat. Mater.* 1994, 3 (4), 337–340. Copyright 1994, with permission from Elsevier. (b) Vertical PACVD reactor utilizing a tungsten filament to assist c-BN growth.[85] Reprinted from Ichinose, Y. *et al.* Synthesis of cubic BN from the gas phase by a new plasma chemical vapour deposition method using r.f. waves and a tungsten filament. *Surf. Coat. Technol.* 1990, 43, 116–127. Copyright 1990, with permission from Elsevier.

resonance (ECR) has also been integrated into PACVD (Fig. 3.5)) to enhance precursor ionization and dissociation and promote collision of these species with the growth substrate, rather than in the plasma.[102] Consequently, large-area films were produced using this technique, compared to the DC jet method.[103]

Temperature was a key experimental parameter in this ECR-assisted PACVD process; temperatures above 1,100 °C led to high-energy F radicals yielding low-quality c-BN films, whereas temperatures below 400–650 °C led to local delamination of the film, while c-BN content decreased below 40%. At 900 °C, the film produced was 40% thicker than 1,000 °C, while clear crystal facets are observed.[104] The use of diamond over Si (001) suppressed the formation of h-BN, with only a top layer of t-BN on the c-BN film being observed *via* Raman spectroscopy.[105] Using a substrate bias of −20 V suppressed the formation of this t-BN layer, with c-BN being formed instead *via* direct epitaxial growth on the diamond.[106] Such epitaxial growth of c-BN had only been observed previously using ion beam-assisted CVD.[107]

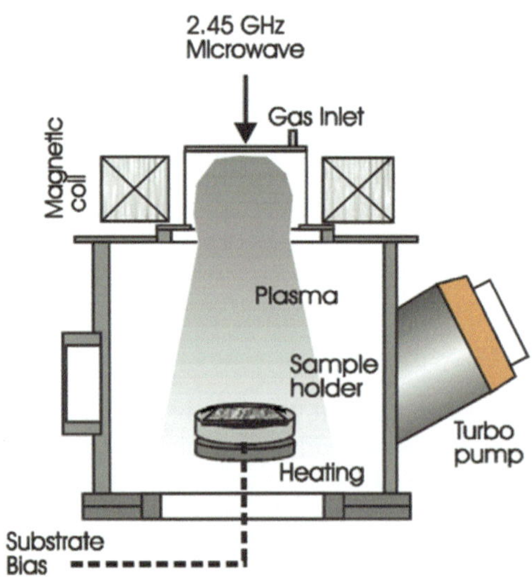

Figure 3.5 Schematic representation of an ECR-enhanced microwave-assisted PACVD system for depositing pure c-BN. The integration of ECR allows for larger area deposition compared to DC jet PACVD at less extreme conditions.[102] Reprinted by permission from Springer Nature Customer Service Center GmbH: Springer Nature. Zhang, W. J. *et al.* Growth behavior of cubic boron nitride films in a two-step process: Changing bias voltage, gas composition, and substrate temperature. *Adv. Funct. Mater.* 2002, 12 (4), 250–254. Copyright 2002.

Figures 3.6 and 3.7 show the progression of c-BN synthesis using PACVD. The SEM micrographs shown in Fig. 3.6 are of c-BN grown from dimethylamine borane with different concentrations of H resulting in varying morphologies. Figure 3.7 shows images of c-BN grown on diamond substrates. Considering the structural similarities between c-BN and diamond, the diamond serves as a natural growth substrate for c-BN as this natural epitaxy suppresses h-BN formation, as shown by Raman spectroscopy. Rather than using diamond-scratched Si, Chong *et al.*[104] used CVD-grown diamond surfaces as substrates and utilized the ECR-enhanced microwave PACVD method to obtain 3 μm thick, pure c-BN films with faceted surfaces (Fig. 3.7c). Faceted surfaces were also observed by Matsumoto *et al.*[95] in films over 20 μm thick films; however, h-BN was still present.

(a) **(b)**

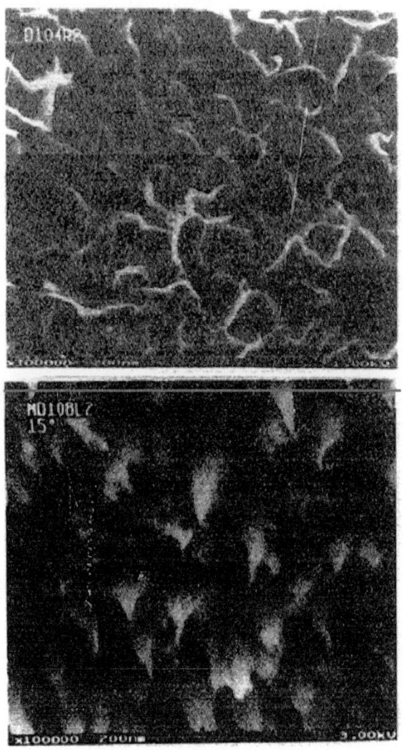

Figure 3.6 Morphology of c-BN films deposited by PACVD at (a) 5.2% and (b) 1.8% dimethylamine borane-hydrogen ratio.[86] Reprinted from Loeffler, J. *et al.* PACVD nano-crystalline BCN thin films obtained by use of an organoboron precursor. *Diamond Relat. Mater.* 1997, 6 (5), 608–611. Copyright 1997, with permission from Elsevier.

3.2.5 c-BN Growth Mechanism

The key growth species during plasma CVD growth of c-BN are atomic H and F present in the reactor. Zhang and Matsumoto have shown that the primary role of F was to etch a-BN and h-BN phases, while H promotes the formation of B-N bonds and hence growth itself.[93] The H:F ratio is a key growth parameter for producing c-BN, as too much H promotes the formation of h-BN. Zhang *et al.*[108] originally studied the growth mechanisms of c-BN on Au during synthesis from BF_3, N_2, and H_2 in He and Ar plasma CVD. The growth precursors BF_x and NH_x drive to the surface-mediated formation of BN, while F etched h-BN, t-BN, and a-BN phases. The incorporation of BF_x at N-terminated sites proved to be slower than that of NH_x at B-terminated sites, thus the surface was B-deficient during growth. N-based clusters dominated the plasma compared to B-based clusters. Subsequently, the F-terminated B atoms on the surface had a lower coverage rate than the H-terminated N atoms. Growth effectively constituted an ion bombardment process, rather than layer-by-layer growth.

The experimental observations of Zhang *et al.*[108] agree with theoretical calculations reported by Larsson *et al.*[109–113] and Mårlid *et al.*[114, 115] who extensively investigated growth processes during c-BN. *Ab initio* adsorption and abstraction energy calculations showed that N(111) and B(111) surfaces can both be stabilized by F termination. The formation of BF_2, however, can hinder the formation of a second N(111) layer *via* blocking growth sites; charge redistribution between B and N further decreases surface reactivity during growth.[116] This means that, during experimental synthesis, F atoms are likely removed by ion bombardment, freeing active sites on the growth surface. Conversely, H was seen as an effective surface-terminating agent; its smaller size enabled adsorption of incoming growth agents during the growth process. Olander *et al.*[117, 118] examined the adsorption energies of B, BBr_x, N, and NH_x species on B and N sites of c-BN(111) surfaces. They showed that N (515 kJ mol^{-1}) and NH_2 (451 kJ mol^{-1}) bound the strongest to the B site with B (531 kJ mol^{-1}) and BBr_2 (479 kJ mol^{-1}) bound the strongest to the N site. The NH_2 migration barrier on B(111) was 41 kJ mol^{-1} and 232 kJ mol^{-1} on N(111) while BBr_2 on N(111) was 43 kJ mol^{-1}

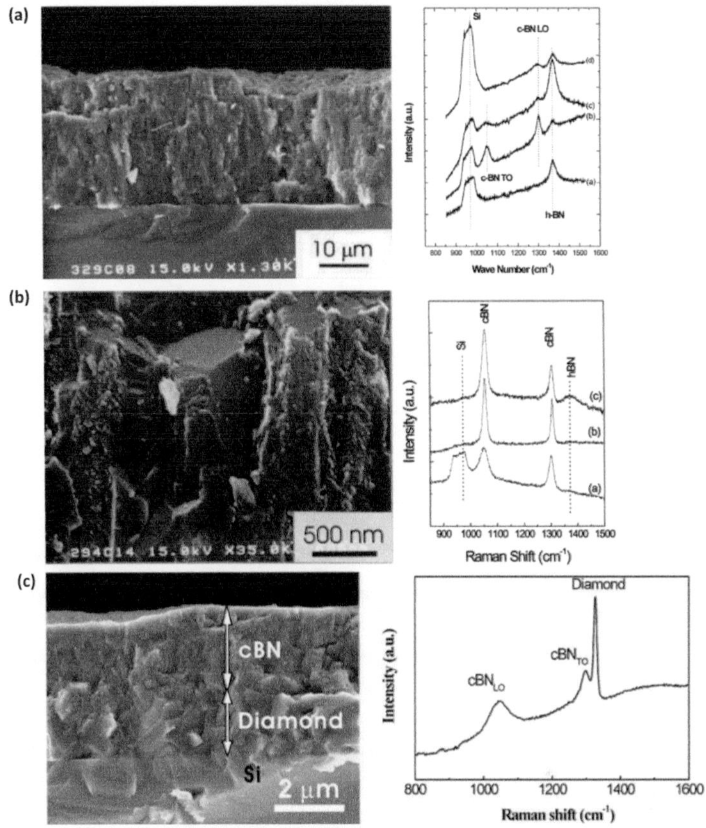

Figure 3.7 (a) SEM micrograph of c-BN containing h-BN, as detected *via* Raman spectroscopy (right), deposited on diamond-scratched Si using RF PACVD.[94] Reprinted by permission from Springer Nature Customer Service Center GmbH: Springer Nature Zhang, W. J. *et al.* The effects of dc bias voltage on the crystal size and crystal quality. of cBN films. *Appl. Phys. A* 2000, 71 (4), 469–472. Copyright 2000. (b) SEM micrograph of c-BN containing h-BN (see Raman spectra inset) deposited in Si(001) by DC jet PACVD.[98] Reprinted from Matsumoto, S. *et al.* The introducing of fluorine into the deposition of BN: a successful method to obtain high-quality, thick cBN films with low residual stress. *Diamond Relat. Mater.* 2001, 10 (9–10), 1868–1874, Copyright 2001, with permission from Elsevier. (c) SEM micrograph of pure c-BN grown on diamond using microwave ECR-enhanced PACVD.[104] Reprinted from Chong, Y. M. *et al.* Growing cubic boron nitride films at different temperatures. *Diamond Relat. Mater.* 2006, 15 (4–8), 1155–1160. Copyright 2006, with permission from Elsevier.

a

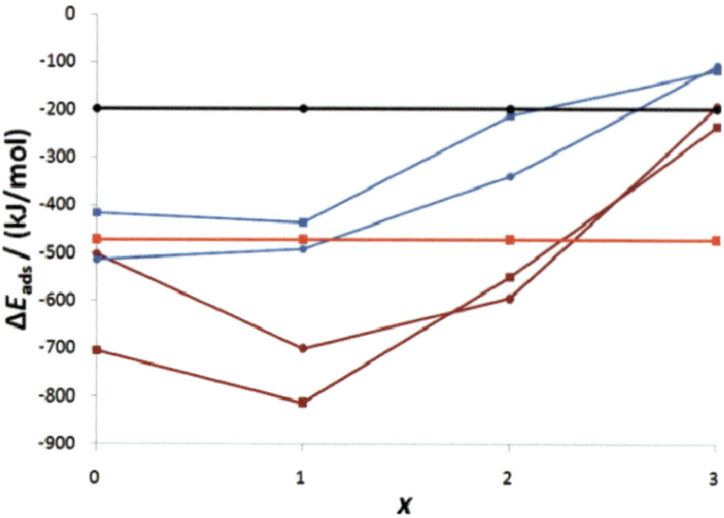

b

Growth species	Gas-phase H	Gas-phase F	Gas-phase H/F
B	-	-	+
BH	+	+	+
BH_2	+	+	+
BH_3	-	-	-
BF	+	+	+
BF_2	+	+	+
BF_3	-	-	-
N	-	+	+
NH	+	+	+
NH_2	+	+	+
NH_3	-	-	-

Figure 3.8 (a) Adsorption energies for the NH_x species (x = 0,1,2,3) on a H-terminated (brown circles), F-covered (brown squares), B-terminated c-BN(100)-(2×1) surface and H-terminated (blue circles), F-covered (blue squares), N-terminated c-BN(100)-(1×1) surface. Adsorption energies for H and F are black circles and red squares respectively. (b) Favorable (green +) and unfavorable (red -) growth species during CVD on c-BN in an H-, F-, or H/F-saturated gas-phase.[120] Reprinted with permission from Karlsson, J. *et al.* Adsorption of growth species on the c-BN(100) surface. *J. Phys. Chem. C* 2011, 115 (34), 16977–16983. Copyright 2011 American Chemical Society.

and 213 kJ mol^{-1} on B(111). For comparison, the barrier for CH_2 migration on a diamond is 52 kJ mol^{-1}.[119] Larsson *et al.*[111] also showed that NH_2 and BH_2 have barriers of 183 and 274 kJ mol^{-1} on the H-terminated c-BN B(111) surface, also less mobile than CH_2 on a diamond. BF_2 and NF_2 were found to adsorb strongly to H-terminated surfaces and BF_2 stronger than NF_2 to F-terminated surfaces, indicating the nature of F species hindering BN growth. A more recent DFT study by Karlsson and Larsson[120] classified the optimal species for growth of c-BN during CVD by calculating the adsorption energies of BH_x, NH_x, and BF_x species on c-BN in gas-phases saturated with H, F, or both H and F. The results indicated that both H and F are required for growth as it strikes a balance between abstraction and exothermic adsorption, and additionally promotes the presence of favorable growth species. The adsorption energies of these species and the favorable growth species for c-BN growth are shown in Fig. 3.8.

3.3 Fabrication of Hexagonal Boron Nitride

Of all BN allotropes, h-BN has arguably seen the most synthetic progress over the last 50 years. Early efforts at CVD synthesis in the 1960s only yielded films with high a-BN content and at the best, semicrystalline h-BN. Now, single crystal, monolayer cm^2 wafers can be produced routinely. CVD is the most effective synthesis method for h-BN production. However, tuning the growth condition is critical to achieve a high-quality product. Here, we classify early attempts at synthesizing BN coatings as attempts at h-BN synthesis, as multi-layered h-BN was the intended product. From these early efforts, we move to discuss a key feature of the modern 2D material landscape, monolayer h-BN.

3.3.1 Chemical Vapor Deposition

CVD methods for depositing BN films at high temperatures were patented as early as 1964. A range of synthesis conditions was stated by Basche who utilized NH_3 and BCl_3 are precursors at high temperatures (1,450–2,300 °C) and low pressures (1–50 Torr), with the choice of parameters crucial to obtain the desired number of layers.[121] Rand and Roberts noted in 1968 that the film deposition rate and layer thickness were both dependent on the growth temperature (600–1,080 °C), carrier gas (H_2, N_2, and He), and the ratio between their NH_3 and B_2H_6 precursors.[122] Films up to 6 nm thick were obtained on a range of substrates, including Si, SiO_2, Ta, Mo, and Ge. In the 1970s studies started to focus on tuning experimental parameters to obtain a higher quality h-BN product. Films were also obtained on substrates of Si and pyrolytic graphite from a range of halogenated precursors such as $B_3Cl_3N_3H_3$,[123] NH_3-BF_3,[124] and BCl_3[125] with NH_3. Numerous studies that followed in the 1980s saw success using halogenated precursors, most commonly with BCl_3 and NH_3 precursors in an atmosphere of Ar and H. Transparent h-BN films were reported by several groups on substrates such as fused Si and sapphire,[126] Cu,[127] graphite,[128–129] and SiO_2.[130] Collectively, these studies highlight the importance of H_2 flow,[128] temperature, and pressure[129, 131, 132] on h-BN film quality. There was indeed no defined set of parameters in this stage of the field that led to pure, high-quality h-BN, and the mechanisms of growth were poorly understood. Figure 3.9 shows how experimental parameters such as deposition temperature and pressure influenced the phase and morphology of h-BN produced from studies in the 1980s. Figure 3.10 shows two different CVD reactor setups. With each substrate used, the morphology and quality of the film were variable. Nodular deposits observed by Takahashi following growth of 100 μm-thick h-BN films on carbon steel led to the hypothesis that the substrate Fe content played an important catalytic role during h-BN growth.[133, 134] A subsequent study provided evidence of catalytic activity using a Ni substrate.[135]

As discussed in Section 3.2, obtaining high-quality c-BN means controlling the deposition of unwanted h-BN, or using ion bombardment during the growth process to etch the h-BN intermediate layer. Throughout the late 1980s to early 1990s, BN films often contained a layer of a-BN next to the substrate, irrespective of whether the target phase was c-BN or h-BN. Amongst the first reports to circumvent this issue was that of Andujar *et al.*,[136] who used RF PACVD, to grow h-BN from B_2H_6 (in H_2) and NH_3 on Si(100). It was proposed that the H etched the a-BN layer during growth. A similar method by Sugino *et al.*[137] produced h-BN from BCl_3 and N_2 with H_2. The same method could produce S-doped h-BN phases, *via* vaporizing solid S into the feedstock gas flow.

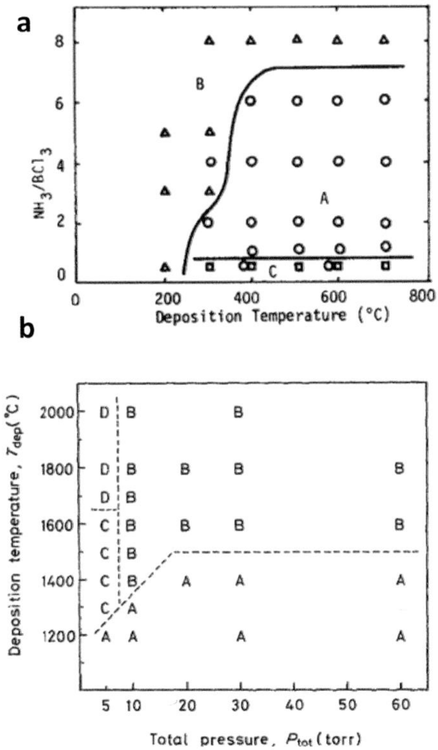

Figure 3.9 (a) Influence of the deposition temperature and gas flow ratio (NH_3:BCl_3) on the morphology of h-BN deposits, thick films (circles), thin films (squares), and white powders (triangles). A, B, and C correspond to thick film, thin film, and powder deposition regions, respectively.[127] Reprinted from Motojima, S. *et al.* Low-temperature deposition of hexagonal BN films by chemical vapour deposition. *Thin Solid Films* 1982, 88 (3), 269–274, Copyright 1982, with permission from Elsevier. (b) Influence of deposition temperature and total pressure on the crystal structure of the BN. A, B, and C correspond to isotropic t-BN, anisotropic t-BN, and a mixture of t-BN and h-BN, respectively.[129] Reprinted by permission from Springer Nature Customer Service Center GmbH: Springer Nature Matsuda, T. *et al.* Synthesis and structure of chemically vapour-deposited boron nitride. *J. Mater. Sci.* 1986, 21 (2), 649–658. Copyright 1986.

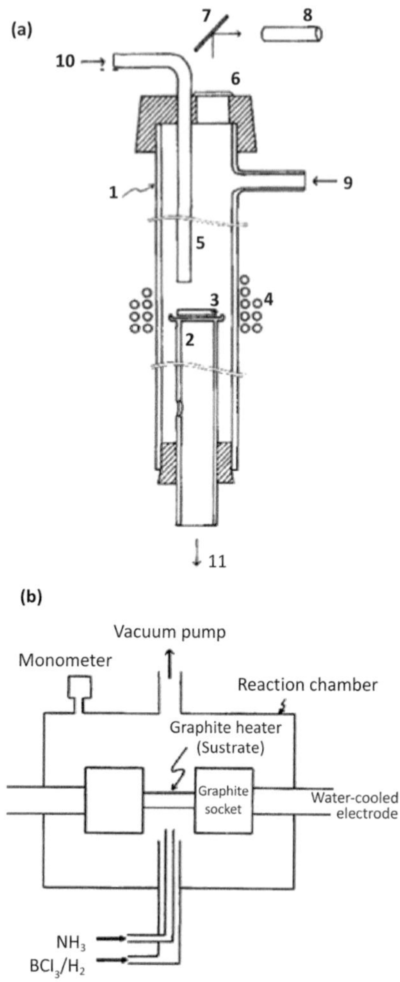

Figure 3.10 (a) Schematic diagram of vertical CVD reactor used by Takahashi, T, *et al.* to obtain h-BN films on Fe. Key labeled parts include the substrate (2), RF coil (4), NH_3 + H_2 inlet (9), BCl_3 + Ar inlet (10) and outlet (11).[133] Reprinted from Takahashi, T. *et al.* Chemical vapor deposition of hexagonal boron nitride thick film on iron. *J. Cryst. Growth* 1979, 47 (2), 245–250, Copyright 1979, with permission from Elsevier. (b) Schematic of horizontal CVD reactor employed by Matsuda *et al.* to grow h-BN on graphite.[129] Reprinted by permission from Springer Nature Customer Service Center GmbH: Springer Nature, Matsuda *et al.* Synthesis and structure of chemically vapour-deposited boron nitride. *J. Mater. Sci.* 1986, 21 (2), 649–658. Copyright 1986.

A shift in focus from h-BN films to the formation of h-BN monolayers occurred in the mid-1990s. Monolayer h-BN was first grown epitaxially by Nagashima *et al.*,[138–140] *via* deposition from borazine on Ni(111), Pd(111), and Pt(111) at 700–800 °C, who noted the weaker interaction with the underlying surface compared to graphene grown under the same conditions.[141] Strained monolayer h-BN was also synthesized on Ni(100) using diborane and ammonia.[142] It is likely that the monolayer strain here arises from the increased lattice mismatch (2.5%) between Ni(100) and h-BN and that of Ni(111) and h-BN (0.4%). Self-limiting h-BN growth was reported by Nagashima *et al.*[140] and by Auwärter *et al.*;[143] in these experiments, once a monolayer of h-BN was formed on the Ni(111) substrate, the growth of a second monolayer became difficult. Auwärter *et al.*[143] in particular observed (scanning tunneling microscope (STM) and X-ray photoelectron diffraction (XPD)) the formation of large triangular monocrystalline h-BN islands occupying either the fcc or hcp sites of Ni(111), with N sitting atop a Ni atom and B sitting in an fcc site. These results agreed well with low-energy electron diffraction (LEED) analysis by Gamou *et al.*[144] and led them to study grain boundary formation during coalescence of islands with different orientations.[145–147] Such grain boundaries persisted during h-BN growth on Ni(111) and Ni(110) from trichloroborazine.[148] Figure 3.11 shows the first observation of the h-BN lattice by Auwärter *et al.* and subsequent observations of the triangular h-BN islands with the proposed structure of the grain boundary (interface of two h-BN islands). Larger h-BN domain sizes were obtained using polycrystalline Ni catalyst foils by Shi *et al.*[149] Here the size of the h-BN domain was only limited by that of the Ni grains, so h-BN grain boundaries were not removed entirely. At the time, further progress in grain boundary reduction was impeded by a lack of understanding regarding h-BN edge termination, how to control h-BN island orientation, and the nature of the boundary between two coalescing islands.

In this respect, the use of Cu surfaces for synthesizing large domain monolayer h-BN grew in popularity following the growth of 2–5 cm^2 layer h-BN films from ammonia borane.[150] The growth mechanism here was clearly surface-mediated.

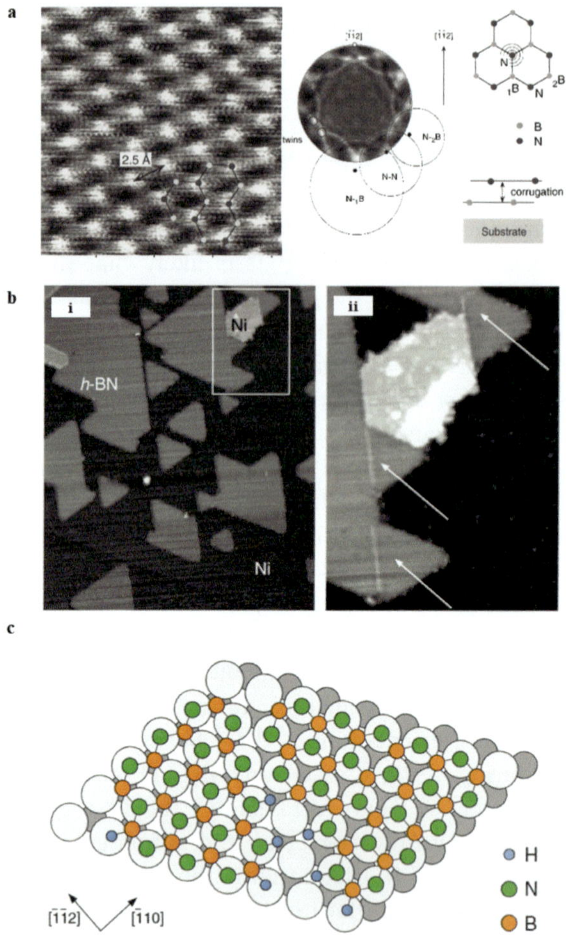

Figure 3.11 (a) (left) STM image of h-BN lattice showing hexagonal honeycomb pattern, spacing 2.5 Å. (right) X-ray photoelectron spectroscopy (XPS) pattern of h-BN on Ni(111) with corresponding honeycomb lattice structure.[143] Reprinted from Auwärter, W. *et al.* XPD and STM investigation of hexagonal boron nitride on Ni(111). *Surf. Sci.* 1999, 429 (1–3), 229–236, Copyright 1999, with permission from Elsevier. (b) (left) STM image of triangular h-BN islands on Ni(111). (right) Highlighted area at higher magnification with arrows pointing to grain boundaries between two oppositely oriented h-BN islands. (c) Structural model of the h-BN/Ni(111) grain boundary unable to coalesce due to N-terminated edges (green) and H-terminated edges (blue).[145] Reprinted from Auwärter, W. *et al.* Defect lines and two-domain structure of hexagonal boron nitride films on Ni(111). *Surf. Sci.* 2003, 545 (1–2), L735–L740, Copyright 2003, with permission from Elsevier.

Beyond this fact, little was understood regarding the h-BN growth mechanism on Cu, aside from the fact that it interacts comparably weaker with h-BN than Ni.[151] Like Ni, the lattice mismatch between h-BN and Cu is low. As such, crystalline BN nanosheets could be grown epitaxially on the surface.[152–154] Also like Ni, the morphology of the Cu surface affected the location and density of h-BN nucleation. Most often, triangular domains were observed prior to coalescing into a monolayer comprising grain boundaries. Kim *et al.*[154] found that tuning the temperature at the ammonia borane precursor was sublimated altered the morphology of the final product (Fig. 3.12). Tuning other experimental parameters such as growth time and pressure enabled lower growth rates, and thus a reduction in grain boundary defects.[155, 156]

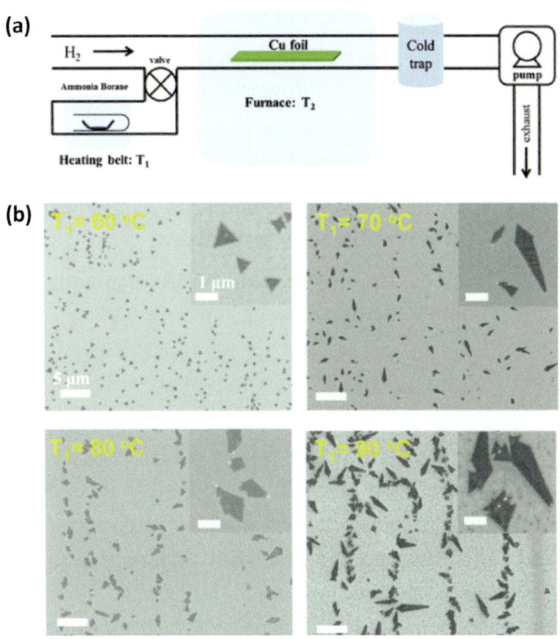

Figure 3.12 (a) Schematic of the CVD reactor setup used by Kim *et al.* to synthesize h-BN under low pressure on Cu foil. T1 refers to the temperature at the ammonia borane precursor is sublimated while T2 is the growth temperature. (b) SEM images illustrating the dependence of the ammonia borane sublimation temperature on the morphology of the h-BN islands on Cu.[154] Reprinted with permission from Kim, K. K. *et al.* Synthesis of monolayer hexagonal boron nitride on Cu foil using chemical vapor deposition. *Nano Lett.* 2012, 12 (1), 161–166. Copyright 2012 American Chemical Society.

Using an ammonia borane precursor, Tay *et al.*[157] systematically studied the effects of growth temperature, precursor temperature, precursor flow rate, substrate position, and growth time in this respect. The density of the film and the h-BN domain size increased as the substrate was placed further downstream in the reactor. For a given position in the reactor, the h-BN domain density increased with growth time, but decreased with too much precursor. Increasing the temperature or growth time also increased the h-BN domain size. Results are summarized in Table 3.1.

Table 3.1 Factors influencing h-BN film growth on Cu from ammonia borane[157]

CVD Parameter	Effect on h-BN Film Growth
Substrate Position	Density and h-BN domain size increase as the substrate is moved further downstream
Growth Temperature	Growth temperature above 1,000 °C is required for high-quality h-BN films
Growth Duration	Density, number of domains, grain size, and film thickness increase with longer growth time
Precursor Temperature	Heating the temperature to 60 °C provides the largest domain size, decreasing above 60 °C
Amount of Precursor	Density and film thickness increase with more precursor, however, domain sizes decrease

Wang *et al.*[158] utilized the same CVD growth process to grow few-layer h-BN films on Cu, while Stehle *et al.*[159] observed that changing the substrate placed in the reactor determined the h-BN island morphology (e.g., from triangles to truncated triangles with concaved edges and hexagons) as the distance between the substrate and the precursor increased (Fig. 3.13). This morphological dependence was attributed to increased B incorporation into the Cu surface at positions further downstream in the reactor. This was consistent with experiments of Kidambi *et al.*[160] who used polycrystalline Cu to grow h-BN from borazine. Here, again the h-BN growth mechanism was surface-mediated with X-ray techniques revealing a lattice expansion of the Cu substrate that was attributed to B incorporation. Tay *et al.*,[161] using a triethylamine borane precursor, achieved monocrystalline monolayer h-BN on Cu, noting that the sublimation temperature of the precursor greatly influenced the quality of the h-BN; this temperature is obviously precursor-specific, 40 °C is the reported threshold for monocrystalline h-BN using triethylamine borane, whereas 60 °C gives the best results for ammonia borane.[157] CVD reactor pressure is also key in this respect; high pressures yield sp^3-hybridized BN due to incomplete dissociation of the (ammonia borane) precursor.[162] Experiments show that under unbalanced conditions, ammonia borane readily polymerises in the gas-phase leading to poor growth rather than dissociation and diffusing on the Cu surface.[163] This is consistent with observations reported from MD simulations.[164]

In 2015, a breakthrough was observed in the growth of h-BN on Cu by Lu *et al.*[165] by using a Cu-Ni alloy. Single-grains of h-BN up to 7,500 μm^2 were obtained using this approach—two orders of magnitude larger than the record grain size at the time. The inclusion of 30% Ni into the Cu catalyst decreased the nucleation density significantly, with the decomposition of ammonia borane assisted by the formation of Ni-B and N-N phases.[166] All h-BN grains were triangular and terminated with N-terminated ZZ (N-ZZ) edges that have the ability to grow over the alloy grain boundaries (Fig. 3.14a). A similar catalyst was also used to produce high-quality h-BN/graphene heterostructures.[167] Subsequently, Ni-Fe alloys also enabled the synthesis of large-area h-BN monolayers, owing to the ability to tune B and N solubilities by varying the alloy composition.[168] The use of Cu as a substrate for h-BN has now become widespread, as it enables single-crystal grains can be grown on the micrometer scale, while also supporting additional growth strategies such as surface oxidation, nanoparticle formation, and water-assisted growth.[169–175] Song *et al.*,[176] for example, grew h-BN monolayer domains 72 μm in length on Cu(111) that aligned along the high symmetry facet. Domain alignment is responsible for the largest single-crystal h-BN structure synthesized to date, 100 cm^2, is reported by Wang *et al.*[177] who used a Cu(110) vicinal surface for h-BN growth, obtained by annealing an industrial Cu foil (Fig. 3.14b). While highly aligned h-BN domains can be formed over large areas on

Ni(111),[175, 178] antiparallel domains can nucleate making it harder to achieve single-crystal of h-BN beyond the micrometer scale.

The intense investigation into Ni and Cu as CVD catalysts for h-BN synthesis since the 1960s has not precluded the study of other transition metals for the same purpose. Catalysts with large lattice mismatch to h-BN, such as Rh(111) (6.7%),[179, 180] Ru(0001) (8.2%),[181] Pd(111) (9%),[182] and Pd(110) (9%),[183] are notable examples in this respect. The first study of h-BN synthesis on Rh(111) in 2004 revealed the formation of a highly regular "nanomesh" with 3 nm periodicity and 2 nm hole size, completely covering the surface as a bilayer.[179] Very similar h-BN nanomeshes have been observed on Ru(0001) (Fig. 3.15).[181] The 2 nm holes were formed by regions where the h-BN binds the strongest to the surface.[180] Incorporating Rh(111) into a multilayer system with YSZ-Si(111) (yttrium stabilized zirconia) also saw the formation of a larger nanomesh superstructure[184] that could be regenerated following oxidation.[185] Lattice mismatch between h-BN and the substrate is not the sole requirement for the formation of these nanomeshes, Pd(111)[182] enables the synthesis of closed monolayer h-BN, despite a mismatch of 9%. h-BN nanomeshes are instead the result of repeating domains in which the h-BN bilayer interacts more strongly with the underlying substrate, giving rise to corrugation in the structure and the appearance of 'holes' in its structure.[180] Each hole is surrounded by domains, which interact relatively weakly with the substrate. For instance, Preobrajenski *et al.*[186] cited specifically the stronger orbital hybridization on Rh(111) results in a stronger attraction of h-BN to specific sites, resulting in a nanomesh. This explanation accounts for the fact that nanomeshes have not been reported on weaker interacting catalysts such as Pd(111),[182] Ag(111),[187] and Pd(110).[183] Similarly, the weak interaction between h-BN and Pt(111) also saw the formation of near-flat monolayers using precursors such as trichloroborazine,[188] borazine,[186] and ammonia borane[189, 190] on the catalyst surface as unique domain structures merged together. A benefit of weak-interaction catalysts, such as Pt(111), however, is that the h-BN grown could be easily transferred from the catalyst substrate, allowing for the Pt foil to be reused for synthesis.[189, 190] The growth mechanism of h-BN on such catalyst is most likely surface-mediated, in line with that observed on Cu, following experiments by Park *et al.* on Pt(111) surfaces.[191] The morphology of the underlying surface also plays an important role, as Pt(001) produced much thinner films compared to Pt(111).[191] The thickness of the film could be better controlled with deposition time in a plasma-assisted method.[192]

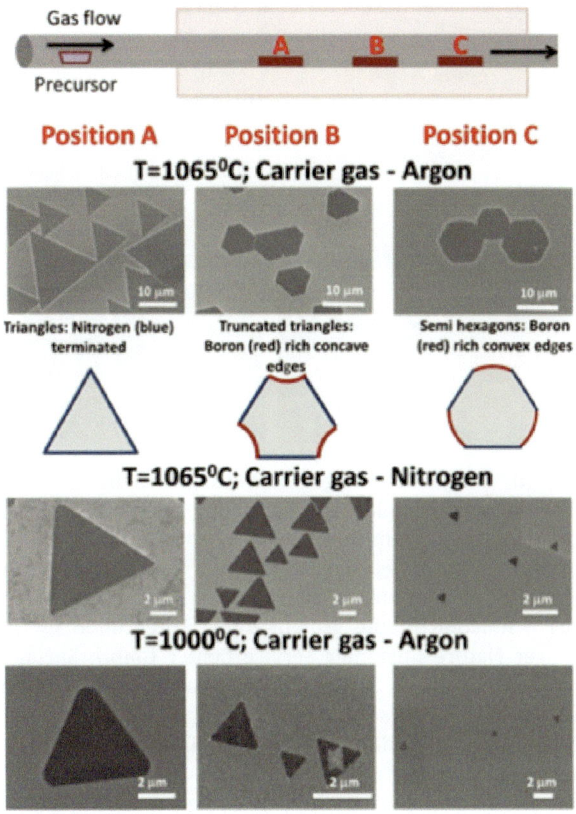

Figure 3.13 Schematic showing different positions, A, B, and C of the growth substrate in the CVD reactor for the synthesis of h-BN by Stehle, Y. *et al.* on Cu. Depending on the position, temperature, and carrier gas the morphology, concentration of the h-BN is controlled.[159] Reprinted with permission from Stehle, Y. *et al.* Synthesis of hexagonal boron nitride monolayer: Control of nucleation and crystal morphology. *Chem. Mater.* 2015, 27 (23), 8041–8047. Copyright 2015 American Chemical Society.

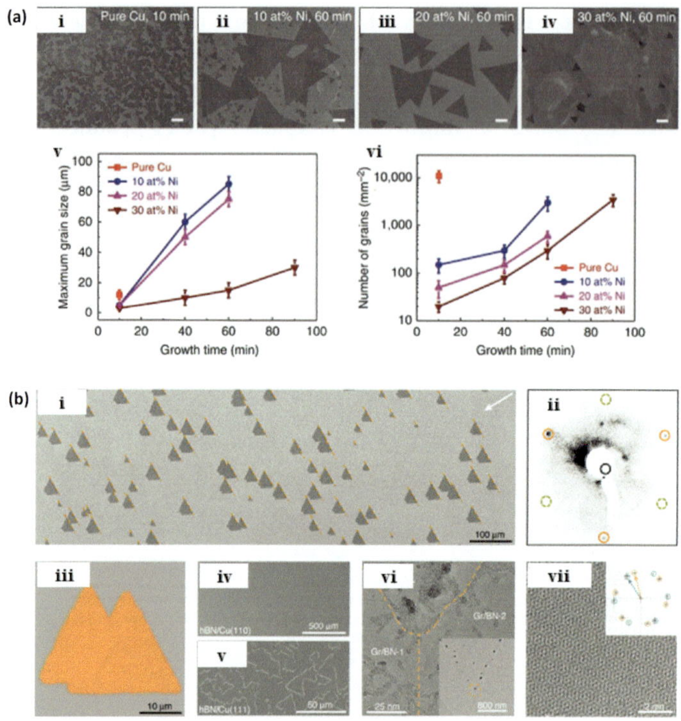

Figure 3.14 (a) (i-iv) SEM images of h-BN grown by CVD on Cu with varying amounts of Ni alloying, displaying significant changes in the nucleation density of h-BN islands. (v) Dependence of the maximum grain size observed on each catalyst as a function of growth time. (vi) Dependence of the number of grains per square millimeter for each catalyst as a function of the growth time.[165] Reprinted by permission from Springer Nature Customer Service Center GmbH: Springer Nature Lu, G. *et al.* Synthesis of large single-crystal hexagonal boron nitride grains on Cu–Ni alloy. *Nat. Commun.* 2015, 6, 6160. Copyright 2015. (b) (i) SEM image of unidirectional h-BN islands grown *via* CVD on Cu(110). (ii) LEED pattern of as grown h-BN. (iii) Second-harmonic generation (SHG) mapping showing no boundary line between aligned h-BN islands. (iv-v) SEM images of the h-BN (iv) and of the Cu(v) after annealing with H_2 showing boundaries only on Cu, not h-BN. (vi) TEM image at the corner in the area of confluence of the h-BN on graphene support. (vii) high resolution transmission electron microscope (HRTEM) image showing a uniform moiré pattern at the corner highlighted in (vi).[177] Reprinted by permission from Springer Nature Customer Service Center GmbH: Springer Nature Wang, L. *et al.* Epitaxial growth of a 100-square-centimetre single-crystal hexagonal boron nitride monolayer on copper. *Nature* 2019, 570 (7759), 91–95. Copyright 2019.

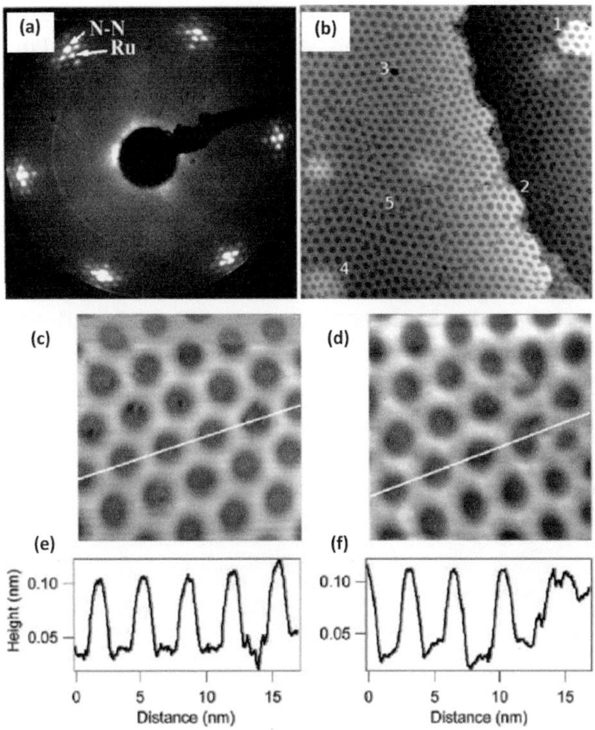

Figure 3.15 (i) LEED pattern of h-BN nanomesh on Ru(0001). (ii) STM image of h-BN nanomesh on Ru(0001) showing (1) single-layer and (2) double-layer step edges, (3) dark pits, (4) surface protrusions, and (5) h-BN nanomesh grain boundaries. (iii) high resolution scanning tunneling microscope (HRSTM) image of h-BN nanomesh on Rh(111) with height cross-section in (v). (iv) HRSTM image of h-BN nanomesh on Ru(0001) with height cross-section in (vi).[181] Reprinted with permission from Goriachko, A. *et al.* Self-assembly of a hexagonal boron nitride nanomesh on Ru(0001). *Langmuir* 2007, 23 (6), 2928–2931. Copyright 2007 American Chemical Society.

3.3.2 h-BN Growth Mechanism

The synthesis of h-BN was initially focused on reducing the c-BN content within a film on disordered, polycrystalline substrates. As this evolved into producing pristine monolayers on single-crystal metal facets, the need to understand the system on a fundamental, atomic scale became more important. Investigation into the metal-adsorbed structure of h-BN, how initial growth precursors decomposed on the catalyst, and the stability of various transition structures (i.e., h-BN islands, edge terminations, etc.) became prominent. Many of the key insights in this respect have been provided by computational studies. Following early studies showing monolayer h-BN growth on Ni(111),[138, 143, 146] Grad *et al.*[193] used DFT to analyze the geometric and electronic structure of h-BN/Ni(111) interfaces as a function of the h-BN adsorption epitaxy. The most favorable adsorption corresponded to N adsorbing atop Ni surface atoms, while B was aligned to the fcc hollow position (Fig. 3.16). However, the energy differences observed were within k_BT for growth temperatures above 1,000 K, indicating kinetic effects may play a key role in h-BN structure determination during growth. These results were consistent with those of Che *et al.*[194] and adsorption energies reported by Huda *et al.*[195] and by Sun *et al.*[196]

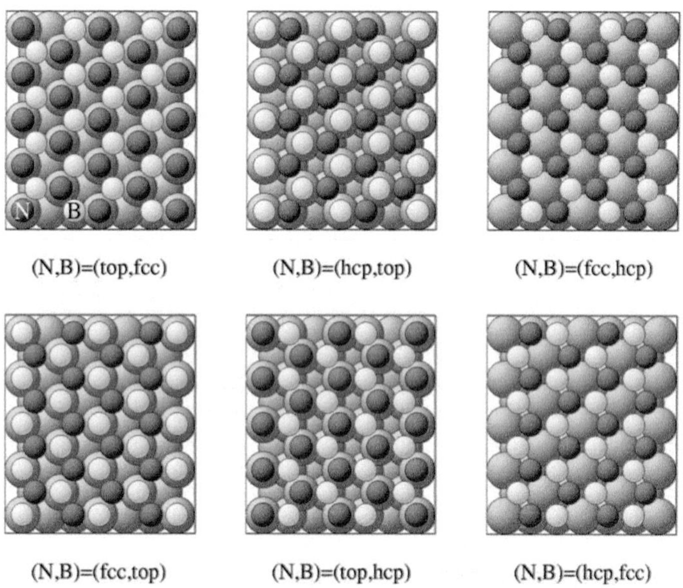

Figure 3.16 Possible (1×1) configurations for monolayer h-BN adsorbed on Ni(111) considering the positions of the B and N atoms relative to Ni.[193] Reprinted with permission from Grad, G. B. *et al.* Density functional theory investigation of the geometric and spintronic structure of h-BN/Ni(111) in view of photoemission and STM experiments. *Phys. Rev. B* 2003, 68 (8), 085404. Copyright 2003 by the American Physical Society.

On Cu(100), Liu *et al.*[197] observed four different orientations of h-BN on the surface and used DFT to calculate the energy of BN clusters on Cu(100) as a function of the rotation angle. The BN clusters exhibited a domed structure, indicating edge interaction with the metal was important. The alignment of one edge of the cluster with the substrate lattice was shown to be the most favorable orientation. BN was observed to not induce significant Cu reconstruction, compared to that induced by graphene monolayers. Alignment of h-BN was also shown to be important on Cu(111) by Song *et al.*[176] as the binding of h-BN was more stable when one edge is parallel to the close-packing direction of the Cu surface (Fig. 3.17).

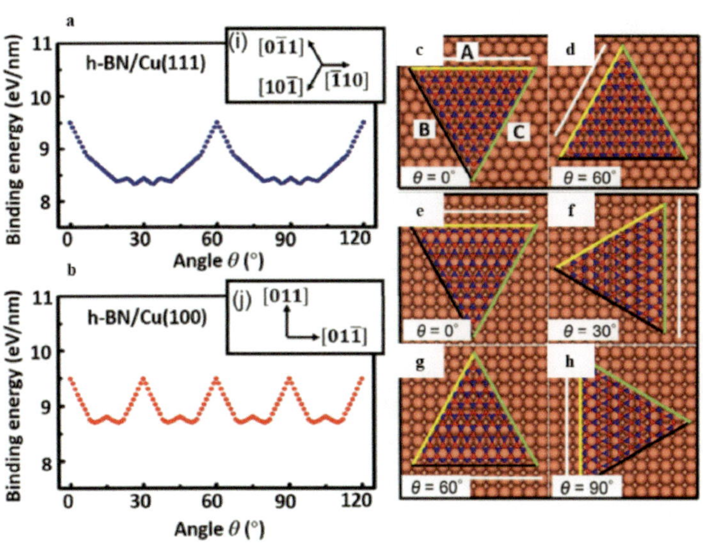

Figure 3.17 DFT calculations of h-BN binding energies on (i) Cu(111) and (ii) Cu(100). (iii-viii) show schematics of h-BN orientation corresponding to the maximum binding energy on (iii-iv) Cu(111) and (v-viii) Cu(100).[176] Reprinted by permission from Springer Nature Customer Service Center GmbH: Springer Nature Song, X. *et al.* Chemical vapor deposition growth of large-scale hexagonal boron nitride with controllable orientation. *Nano Res.* 2015, 8 (10), 3164–3176. Copyright 2015.

Laskowski *et al.*[198] employed an *ab initio*-derived force fields and DFT to examine the starkly different and much larger h-BN nanomesh structures formed on Rh(111). The model proposed disagreed with that originally proposed to explain the origin of h-BN nanomesh,[179] but explained the σ band splitting observed experimentally and was also consistent with other STM work at the time, showing slight Rh corrugation in the top layer. Laskowski *et al.*[199] showed that h-BN interacts strongly with Rh(111) and Ru(001) and weakly with Pt(111), both exhibiting larger corrugations than Pt(111). The Rh(111) exhibited a larger correlation than Ru(001), however, the size of the bonding region to the substrate was larger for Ru(001), indicating strain within individual B-N bonds in the adsorbed h-BN structure (Fig. 3.18). Similar calculations[200] also predicted that Ag(111) and h-BN interacted very weakly hence resulting in no preferred orientation of h-BN islands on the catalyst, despite a large lattice mismatch.[187]

A number of studies have examined the relative edge stabilities on transition metal catalysts pertinent to h-BN island growth. The pristine and H-terminated equilibrium edge energies of h-BN were calculated by Liu *et al.*[201] as a function of the boron chemical potential, μ_B. A similar study was reported by Zhao *et al.*[202] Fu *et al.*[203] analyzed the energetics of triangular h-BN islands using energy decomposition analysis based on density functional tight binding (DFTB) calculations. For free-standing h-BN, Liu *et al.*[201] and Fu *et al.*[203] found that N-terminated ZZ edges were the most stable, while Zhao *et* al.[204] found the armchair edge to be the most stable pristine edge. This changes markedly when the h-BN island is adsorbed to a metal surface, and can be tuned by the metal as shown by Laskowski *et al.*,[200] as *ab initio* DFT calculations showed trends across the periodic table regarding binding energy decreases with the filling of the *d* shell. Zhao *et al.*[204] demonstrated that ZZB+N and ZZN+B Klein edges (Fig. 3.19) are the most stable on Ni(111), Cu(111), and Rh(111) and showed that the predominant equilibrium edge structure and island shape evolved as a function of the difference in B and N chemical potentials ($\Delta\mu$, Fig. 3.19). Zhu *et al.*[205] also studied geometries of h-BN clusters on Ni(111) and Cu(111) predicted the existence of ultra-stable magic clusters on both surfaces.

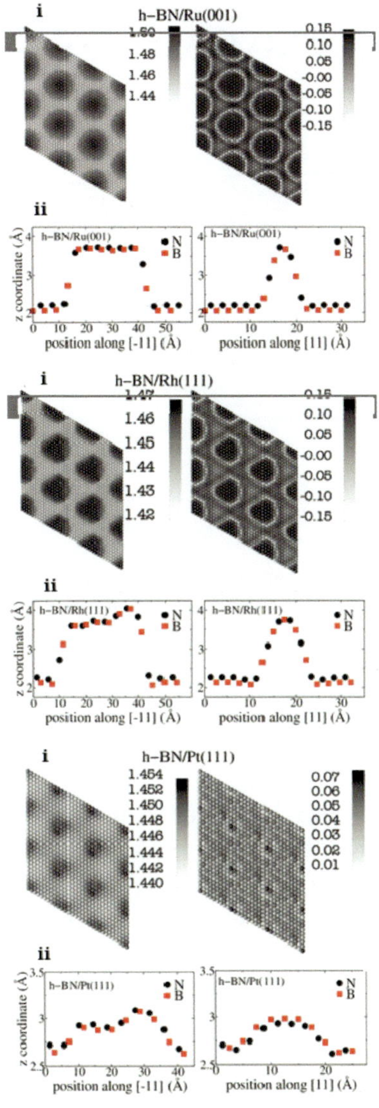

Figure 3.18 (i) Average B-N bond length (left) and buckling (right) in the N-B_3 unit for the (a) h-BN/Ru(001), (b) h-BN/Rh(111), and (c) h-BN/Pt(111) nanomesh.[199] Reprinted with permission from Laskowski, R. *et al. ab initio* study of h-BN nanomeshes on Ru(001), Rh(111), and Pt(111). *Phys. Rev. B* 2010, 81 (7), 075418. Copyright 2010 by the American Physical Society.

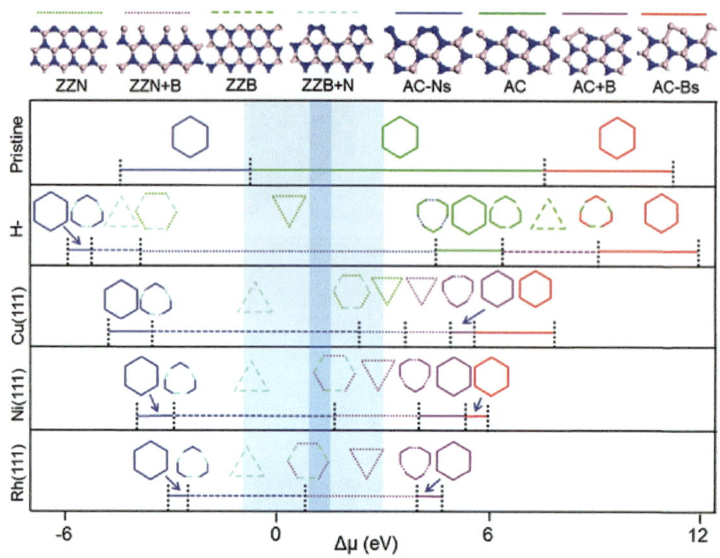

Figure 3.19 Equilibrium shapes and edges structures of h-BN domains pristine, H-terminated, and adsorbed to Cu(111), Ni(111), and Rh(111) as a function of chemical potential difference.[204] Zhao, R. *et al.* The transition metal surface passivated edges of hexagonal boron nitride (h-BN) and the mechanism of h-BN's chemical vapor deposition (CVD) growth. *Phys. Chem. Chem. Phys.* 2015, 17 (43), 29327–29334. Published by the PCCP Owner Societies.

Zhao *et al.*[204] also presented a mechanism to explain the growth kinetics of BN edges (Fig. 3.19). In essence, forming a new ZZ chain requires two steps: forming a hexagon by adding three atoms and the repeatable addition of a BN unit to form new hexagons. The reconstructed ZZB+N/ZZN+B edges require only the addition of a single BN unit. An atomistic growth mechanism for h-BN growth on Cu(111) and Ni(111) was proposed by Zhang *et al.*[206] that also examined the influence of μ as well as temperature and pressure. On Ni(111), results agreed with experimental observations that triangular h-BN islands are the most stable, enclosed by ZZN edges at low μ_B, ZZB at zero μ_B (in contrast to prior experimental observations), and kinked N-terminated edges at high μ_B. On Cu, N-terminated triangles dominate at low and zero μ_B while high μ_B favors a hexagonal h-BN island with alternating ZZB and ZZN edges. This kinetic model revealed optimal atomic configurations during the docking of individual B and N atoms to the growing edge. Interestingly, the kinetics of edge growth revealed that the growth was highly tunable by chemical potential, i.e., the feedstock. The free-energy barriers of adding rows of BN to principal h-BN edges, as shown in Fig. 3.20, were different and monotonically decrease with increasing μ_B as well as the difference between the free energy barriers. This shows the importance of controlling μ_B in controlling the rate of nucleation. In addition, the sequence of edge growth changed with μ_B.

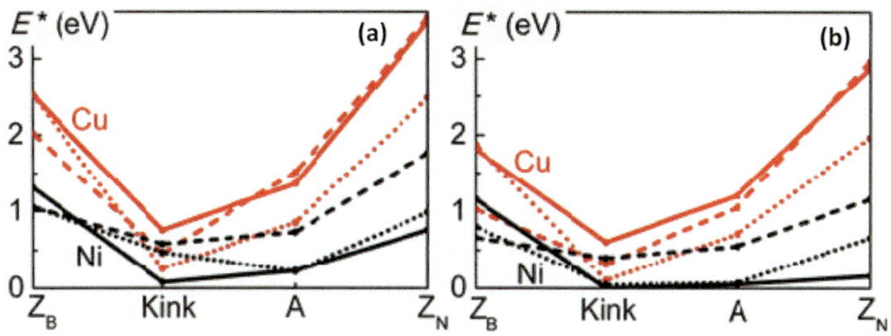

Figure 3.20 (a) Free energy barriers for nucleating BN rows at principal h-BN edges on Ni (black) and Cu (red) with $\Delta\mu_{BN}$ = 0.01 eV. Dashed, solid and dotted lines indicate μ_B = −0.5, 0 and 0.5 eV, respectively. (b) As (a) though $\Delta\mu_{BN}$ = 0.3 eV.[206] Reprinted with permission from Zhang, Z. *et al.* Growth mechanism and morphology of hexagonal boron nitride. *Nano Lett.* 2016, 16 (2), 1398–1403. Copyright 2016 American Chemical Society.

Liu *et al.*[207] used reactive MD simulations to investigate the adsorption and diffusion pathways of B_xN_y (x,y = 1,2) on Ni(111) and Ni(211) on the surface, as well as atomic B and N in the subsurface. While B was found to diffuse in the subsurface (which can hinder h-BN growth by distorting the Ni lattice) and on the surface, N diffuses strictly on the substrate surface only (Fig. 3.21). On this basis, it was proposed an atomic growth mechanism for h-BN from atomic B and N comprising the growth of linear BN chains and evolution into branched networks prior to nucleating hexagonal BN networks. In a subsequent study,[208] reactive MD simulations showed that tuning the chemical potentials of B and N can control the obtained h-BN lattices on Ni(111) from atomic B and N. They found that the presence of N plays a more crucial role in the size of h-BN islands obtained on Ni(111) as the formation of N_2 can restrict h-BN growth.

The morphology of h-BN islands transitioned from triangles to hexagons in N-rich conditions. These simulations are more representative of h-BN synthesis from MBE due to the lack of H. Their work agreed with that of Liu *et al.*[209] (Fig. 3.22), indicating that BN chains evolve into the h-BN networks and islands on (111) surfaces. The latter study claimed that B_nN_{n-1} or $B_{n-1}N_n$ chains were required to evolve into h-BN islands, dominating the minimum energy structures for clusters below a critical size (13 atoms) required for BN network formation.

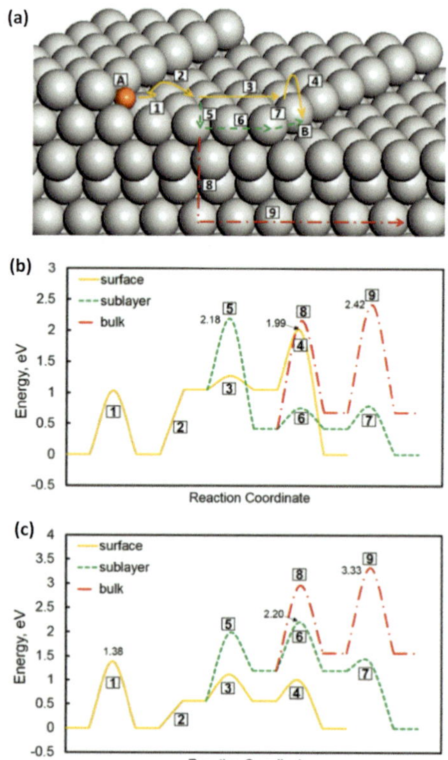

Figure 3.21 (a) Schematic showing atomic B and N pathways in the surface (yellow), sublayer (green), and bulk (red). Gray and orange spheres are Ni and the diffusing atom respectively. Potential energy surface for B (b) and N (c) diffusing through pathways shown in (a).[207] Reprinted with permission from Liu, S. *et al.* atomistic insights into nucleation and formation of hexagonal boron nitride on nickel from first-principles-based reactive molecular dynamics simulations. *ACS Nano* 2017, 11 (4), 3585–3596. Copyright 2017 American Chemical Society.

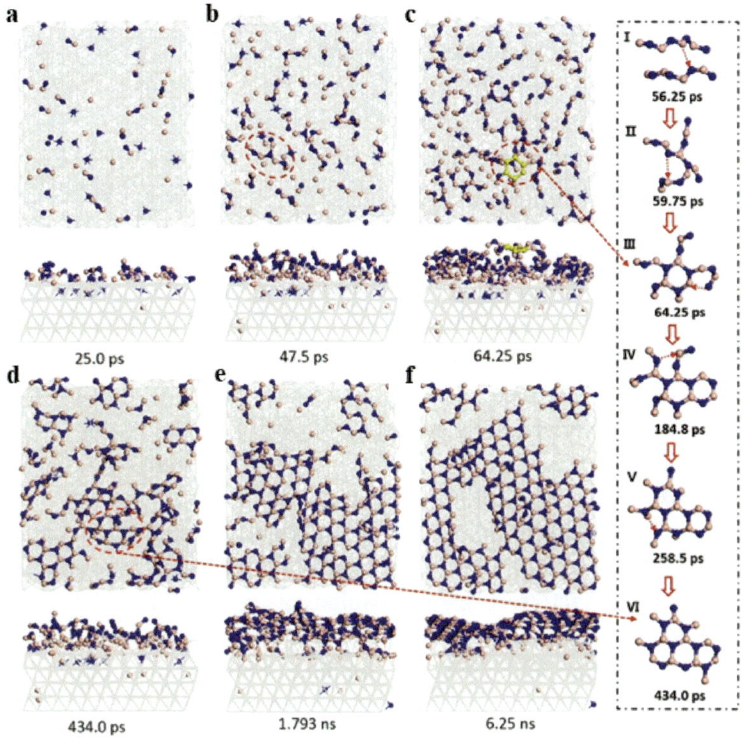

Figure 3.22 Snapshots of reactive MD simulations by Liu, S. *et al.* at 1,300 K during h-BN growth on Ni, showing the evolution of atomic BN to BN chains to BN hexagons and BN ring networks.[207] Reprinted with permission from Liu, S. *et al.* atomistic insights into nucleation and formation of hexagonal boron nitride on nickel from first-principles-based reactive molecular dynamics simulations. *ACS Nano* 2017, 11 (4), 3585–3596. Copyright 2017 American Chemical Society.

3.3.3 Exfoliation

Since the mechanical exfoliation of graphene,[210] the technique was used to pursue pristine flakes of other 2D materials, such as h-BN. Zettl *et al.*[211] studied free-standing h-BN on SiO_2 and characterized them with AFM, TEM, and optical microscopy. While this method does not allow for industrial-level scale-up, it does allow for a high-quality material to be obtained. For example, mechanical exfoliation led to important studies of h-BN structural[212] and optical[213] properties. Recently, mechanical exfoliation of h-BN was performed by shearing in molten Pebax at 200 °C to obtain water-dispersible h-BN/Pebax composites that had attractive antifriction properties for use as a lubricant at low concentrations (0.3 mg mL^{-1}).[214]

A more involved though potentially scalable synthesis is that of liquid exfoliation. Han *et al.*[215] exfoliated h-BN layers from 0.2 mg of h-BN in a solution of dichloroethane and poly(m-phenylenevinylene-co-2,5-dictoxy-p-phenylenevinylene). Organic solvents proved effective media for the exfoliation of h-BN sheets from bulk h-BN,[216–219] sometimes accompanied with a polyethylene-based stabilizing agent to keep the number of h-BN layers low and keep them dispersed.[220, 221] Recent efforts have looked to environmentally friendly materials to assist in exfoliation.[222, 223] A wide range of organic solvents was investigated by Coleman *et al.*[224] for the exfoliation of h-BN, MoS_2, WS_2, and other TMDs showing that pyrol-based compounds and dimethyl sulfoxide (DMSO) were the most effective. Alcohols are also a commonly used and effective organic exfoliating agent. Zhou *et al.*[225] exfoliated h-BN, MoS_2, and WS_2 in a mixture of ethanol in water, with 55 vol% achieving the highest concentration of h-BN. Ammonia has been added to exfoliate this material and appeared to play an important role.[226] Recent work has also shown that the exfoliation efficiency can be tuned with the ratio of alcohol to water.[227] Shen *et al.* considered surface tension components and predicted effective solvents for h-BN and TMD materials, finding that for h-BN, a 1:1 mixture of water and isopropyl alcohol (IPA) is most effective for achieving mono- and few-layer sheets consistently (Fig. 3.23a).[228] The addition of salts to the sonication media has also proven effective in exfoliating h-BN layers with evidence of specific ion effects controlling the degrees of exfoliation and number of h-BN layers.[229, 230] Lin *et al.*[231] achieved exfoliation of h-BN using sonication in water without surfactants or additional organics (Fig. 3.23b). A recent MD study by Mukhopadhyay *et al.*[232] has shown that the polarity of the exfoliating agent changes the mechanism of exfoliation and can potentially explain the observed experimental results regarding different concentrations of alcohols, salts, and the effectiveness of different solvents. Polar solvents strongly adsorb on the surface and in between h-BN layers forming quasi-stable states through electrostatic attraction. This reduces binding between layers and promotes the exfoliation of individual h-BN sheets. The smaller the exfoliating agent, the easier the intercalation due to steric effects, hence the enhanced exfoliation when Li^+ is used over Na^+ and K^+ in salt-assisted exfoliation.[230] Mukhopadhyay *et al.*[232] proposed two mechanisms of exfoliation: perpendicular and parallel, determined by the potential of mean force between h-BN sheets as the slipping and vertical distances between them change with intercalation of the exfoliating agent (Fig. 3.24).

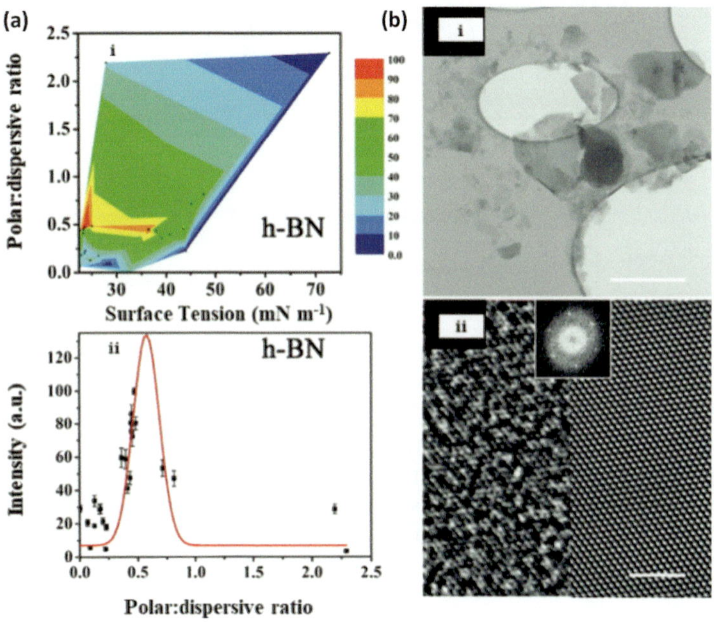

Figure 3.23 (a) (i) Polar dispersive ratio as a function of total surface tension and (ii) fit with Lorentzian peak functions.[228] Reprinted with permission from Shen, J. *et al.* liquid phase exfoliation of two-dimensional materials by directly probing and matching surface tension components. *Nano Lett.* 2015, 15 (8), 5449–5454. Copyright 2015 American Chemical Society. (b) (i) Low magnification TEM of h-BN nanosheets in aqueous dispersions and (ii) HRTEM image of a nanosheet with inset simulated fast Fourier transform.[231] Reprinted with permission from Lin, Y. *et al.* aqueous dispersions of few-layered and monolayered hexagonal boron nitride nanosheets from sonication-assisted hydrolysis: Critical role of water. *J. Phys. Chem. C* 2011, 115 (6), 2679–2685. Copyright 2011 American Chemical Society.

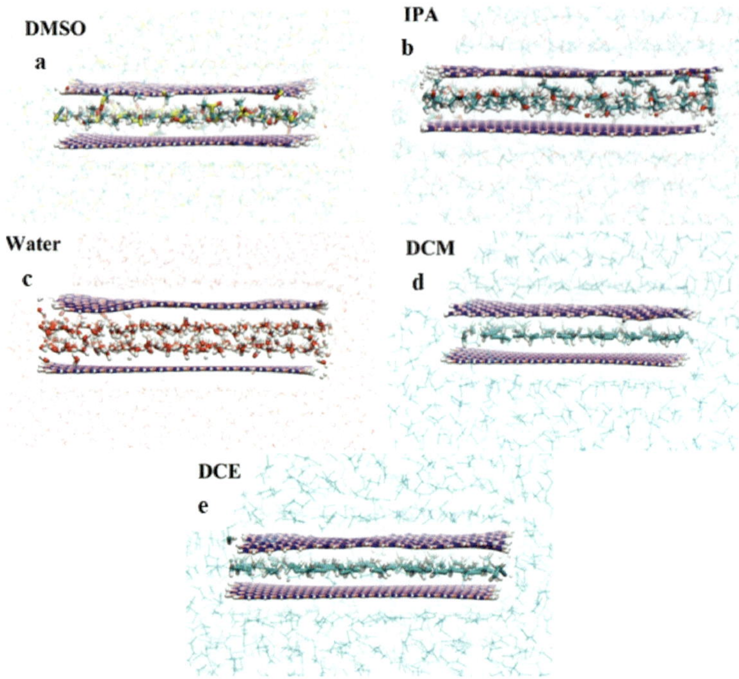

Figure 3.24 Snapshots after 60 ns of exfoliation MD simulations showing (a) DMSO, (b) IPA, (c) Water, (d) dichloromethane (DCM), and (e) DCM intercalated between two h-BN nanosheets.[232] Reprinted with permission from Mukhopadhyay, T. K. *et al.* Deciphering the role of solvents in the liquid phase exfoliation of hexagonal boron nitride: A molecular dynamics simulation study. *J. Phys. Chem. C* 2017, 121 (1), 811–822. Copyright 2017 American Chemical Society.

Wang *et al.*[233] fabricated few-layer h-BN nanosheets *via* a unique process of "chemical blowing", akin to chemical exfoliation. Large bubbles are produced with atomically thin B-N-H polymer walls by H_2 gas from an ammonia borane precursor. Following the collapse of the polymer bubbles and high-temperature annealing, large-area (100 μm) h-BN sheets approximately 2 nm thick were observed. Lee *et al.*[234] exfoliated hydroxylated h-BN nanoplatelets *via* a ball milling method in NaOH seeing yields of 18%.

Traditionally and most frequently, sonication in the presence of the exfoliating agent and seldom stabilizing agents are used for liquid exfoliation. Additional methods have been applied to enhance exfoliation however including high-pressure fluid processing,[225] microwave irradiation,[235,236] thermal treatment,[237–240] and the application of electric fields.[241] Magnetic stirring during sonication has also been shown to alleviate phenomena of h-BN layers adhering together despite adsorption and intercalation, resulting in few-layer h-BN sheets less than 1 nm thick.[242] Post-exfoliation, centrifugation has shown to be effective in sorting h-BN sheets by length.[243, 244]

3.3.4 Sputtering

Several attempts at synthesizing crystalline h-BN films have been reported using magnetron sputtering, following earlier sputtering approaches to synthesize c-BN (Section 3.2.3). The latter yielded h-BN underlayers, though only when deposited in pure N_2 compared to a mixture of Ar and N_2.[39] At first, sputtering was only capable of producing mixed c-BN and h-BN phases. For instance, Gissler *et al.*[245] employed a BN target in a mixed Ar/N_2 atmosphere, to obtain sub-1 μm thick layers of such mixed phases on substrates including steel, Si, and sapphire. Similar results were observed by Jäger *et al.*[246] using a method previously used to deposit c-BN.[54] Sutter *et al.* synthesized few-layer h-BN films from a pure B target onto heated Ru(0001) in N_2/Ar with a consistent thickness (Fig. 3.25).[247] The Ar^+ ions here ablate B *via* collision with the targets, which subsequently react with gas-phase N radicals produced by dissociation of N_2 in the plasma. At 850 °C, crystalline h-BN is achieved consistently, while lower temperatures yield phases of r-BN or a-BN. There are conflicting reports regarding the role of feedstock N_2 on BN morphology; Peng *et al.*[248] reported that a mixture of N_2/Ar is required for a stoichiometric, consistent sputtering of a single h-BN layer on Cu(111), whereas Stewart *et al.*[249] demonstrated that the N_2 feedstock component has the negligible influence of the film composition during RF magnetron sputtering growth at 850 °C; crystalline h-BN films in the latter investigation were only obtained on Ni foil rather than sapphire and Si substrates. This result was attributed to the surface roughness of the Ni foil compared to the near atomically smooth sapphire and Si. While not confirmed, this is possibly the reason that Ohta *et al.*[250] saw the deposition of highly crystalline h-BN on AlN-supported substrates, but not on an AlN substrate itself. Nevertheless, Meng *et al.*[251] reported highly aligned, 600 μm-long single-crystal h-BN domains on a highly crystalline Ni(111) surface on a sapphire substrate using similar sputtering approaches.

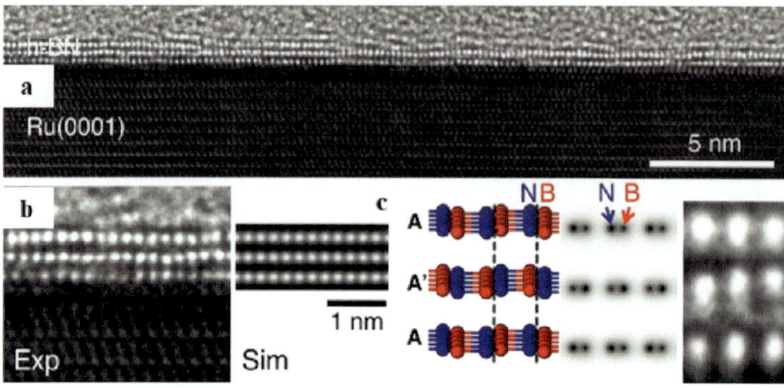

Figure 3.25 (i) HRTEM image of a few-layer h-BN film sputter-deposited into Ru(0001). (ii) Comparison of experimental TEM contrast with dynamic image simulation. (iii) Computed structure and electron density map compared to experimental image assuming A–A–A stacking.[247] Reprinted with permission from Sutter, P. *et al.* scalable synthesis of uniform few-layer hexagonal boron nitride dielectric films. *Nano Lett.* 2013, 13 (1), 276–281. Copyright 2013 American Chemical Society.

3.3.5 Beam Epitaxy

MBE facilitates the efficient growth of monolayer h-BN by offering precise control over growth conditions. In this respect, Ni is by far the most commonly employed catalyst substrate. Previously used to synthesize c-BN,[252] multilayer thick films of h-BN on Ni(111) were produced by MBE by Tasi *et al.*[253] using elemental B evaporated by an electron beam and plasma-generated N at 890 °C. Nakhaie *et al.*[254] later grew monolayer h-BN using MBE on polycrystalline Ni at 730 °C and 835 °C. Here, as the growth temperature increased, h-BN islands transitioned from star-shaped islands to triangles, likely due to the higher growth temperature promoting substrate annealing, leading to fewer imperfections and thus a shift in domain shape during growth. Indeed, a number of independent studies show that h-BN island morphology can be tuned by the crystallinity and smoothness of the surface.[255–257] A subsequent study by Nakhaie *et al.*[258] showed that h-BN growth can proceed overstep edges and grains on Ni(111) at 735 °C. Slower ad-atom diffusion at a lower growth temperature (600 °C), impeded this growth mode, with growth proceeding *via* triangular islands formed with an increased nucleation density. Tonkikh *et al.*[259] have also demonstrated h-BN domains on monocrystalline Ni(111) substrates, with no amorphous or polycrystalline phases observed. Incorporation of C into Ni substrates prior to growth also promotes the formation of single-crystal h-BN flakes,[255] with millimeter-sized flakes reported on Ni(100) surfaces produced from abnormal grain growth.[257] The interstitial carbon in these pre-treated Ni substrates pushes B and N adsorption into regions where fast diffusion is possible, reducing the nucleation activation energy. Co has also proven to be an effective substrate for MBE growth of multi-layer h-BN films.[260, 261] As for Ni, tuning the morphology of the Co substrate *via* polishing enables the growth of h-BN islands with a wide array of shapes and sizes, as shown in Fig. 3.26,[261] as the smoother surface provides a lower diffusion barrier. A rougher surface, by contrast, is accompanied by a high kinetic energy barrier that restricts the long-range diffusion of surface adsorbed growth agents.

Graphite and graphene substrates also enable epitaxial growth of h-BN *via* MBE.[262] For instance, Xu *et al.*[263] first demonstrated h-BN growth on Co-supported graphene, yielding a 2D van der Waals heterostructure. Monolayer h-BN commensurately grown on graphite was achieved by Cho *et al.*[264] and Cheng *et al.*;[265] the degree of alignment between the underlying graphite substrate and adsorbed h-BN monolayer is evidenced by the 6-fold Moiré structure. In a similar vein,[266] diamond (111) and (100) surfaces have been used for epitaxial h-BN growth using triethylboron and NH_3 growth precursors. The diamond (111) surface allows for a highly oriented h-BN layer due to its hexagonal symmetry and a low number of dangling bonds. On the other hand, the diamond (100) facet yielded a disordered h-BN product. Metalorganic precursors were also used to grow h-BN on AlN/sapphire substrates.[267] However, temperatures above 1,000 °C are typically required for crystalline h-BN in these setups.[268]

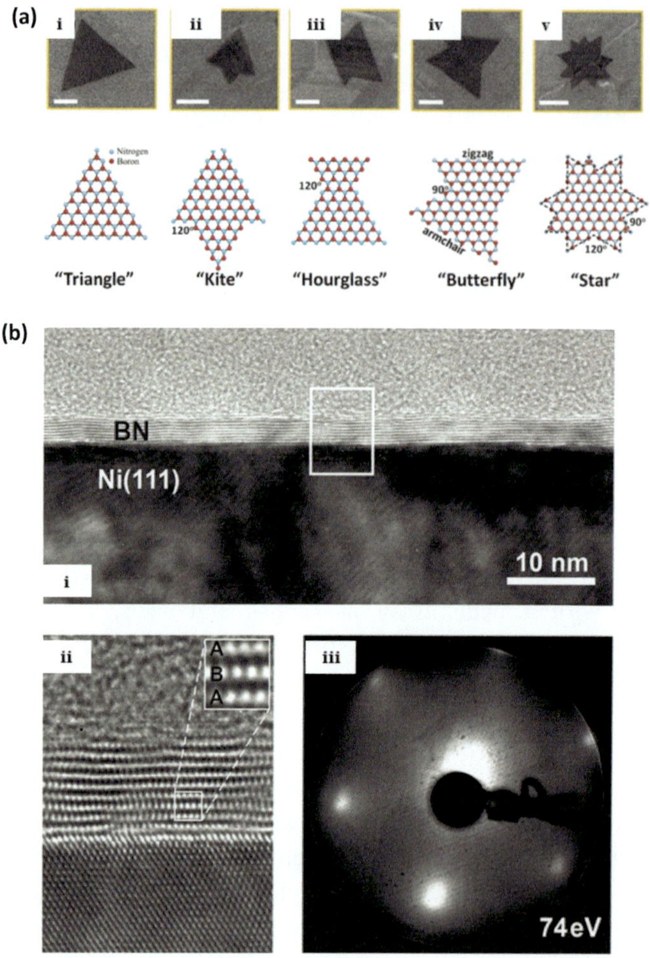

Figure 3.26 (a) SEM images and schematics of h-BN islands observed following MBE on polished Co foils. Scale bar 5 μm.[261] Distributed under a creative commons CC BY license. (b) (i) HRTEM image of a 10-layer h-BN film on Ni(111) grown *via* MBE. (ii) HRTEM showing A–B stacking of BN layers. (iii) LEED pattern following annealing at 400 °C.[259] Distributed under a creative commons CC BY license.

Using so-called flow rate modulated epitaxy (FME), a series of studies by Kobayashi *et al.* demonstrated the epitaxial growth of (0001) h-BN on Ni(111) and sapphire(0001) substrates from triethylboron and ammonia at temperatures over 1,000 °C.[269–271] Heilmann *et al.*[272] grew h-BN epitaxially on graphene using MBE, demonstrating that defects in the graphene substrate, such as step edges and wrinkles, act as h-BN nucleation sites. Interestingly, however, the formation of h-BN nanoparticles was observed near highly strained areas on the graphene substrate. Further, as h-BN islands expand laterally, nanoparticles remain predominantly at the step edges and wrinkles. As h-BN islands of different orientations coalesce, these NPs nucleate further and begin to cluster. The chief challenge with respect to device applications here is suppressing such nanoparticle formation, as these nanoparticles exhibit nearly no tunneling current, compared to pristine h-BN.[272]

3.4 Fabrication of Boron Nitride Nanotubes

The first observation of a "tubular" form of BN was reported by Hamilton in 1993,[273] who heated an a-BN phase at 1,100 °C. These "hollow fibers" were roughly 3 μm in diameter and 50–100 μm in length and were mixed with a t-BN by-product. Similar t-BN nanotubules were observed years later *via* the pyrolysis of trichloroborazine over alumina at 750 °C[274] formed *via* a seemingly similar mechanism to those produced by Hamilton, whereby the growth of tubular BN resulted from corpuscular templates.[275, 276]

BNNTs themselves were first hypothesized by Rubio *et al.*[277] in 1994, only a year after the first definitive images of SWCNTs were reported by Iijima.[278] A wide variety of gas, liquid, and solid phase methods have since been employed to synthesize BNNTs, such as arc discharge, laser ablation, ball milling, substitution, and solution-based routes. The most successful method, however, as is the case for c-BN and h-BN phases, is CVD. This section overviews the key advances reported with respect to each of these synthetic routes to BNNTs.

3.4.1 Arc Discharge

Arc discharge, successful in the synthesis of CNTs[278] and fullerenes,[279] produced the first pure BNNTs in 1995.[280] During arc discharge, a constant current is maintained through two electrodes that come into contact with each other to generate an arc. This arc generates a plasma at high temperatures (~4,000–6,000 K), sublimating the precursors at the anode, where they then aggregate in the gas phase and deposit on the cathode to form the final product at a much lower temperature (see Fig. 3.27).[278, 281] The first BNNTs obtained using arc discharge (W cathode, Cu anode) were multi-walled, exhibited inner diameters of 1–3 nm, and were up to 200 nm in length.

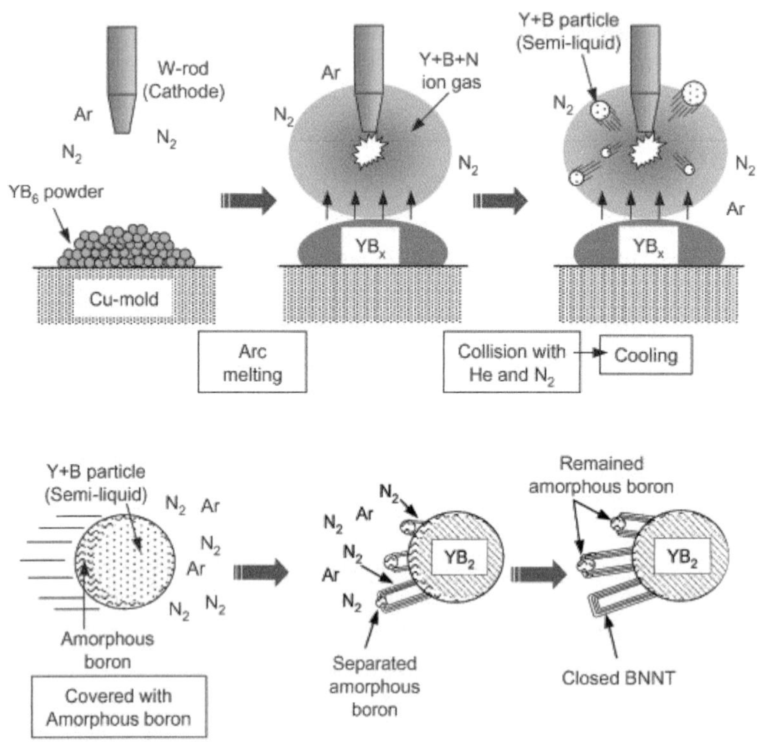

Figure 3.27 Schematic illustration of the formation of BNNTs *via* arc discharge. Ions and radicals of Y, B, and N react in the arc are cooled by colliding with N_2 and Ar to form semi-molten particles of B_xY_y. B presumably precipitates upon cooling and reacts with N in the atmosphere to form BNNTs. Some particles remain capped by amorphous B clusters resulting from precipitation from B_xY_y particles.[281] Reprinted from Narita, I. *et al.* Synthesis of boron nitride nanotubes by using YB6 powder. *Solid State Commun.* 2002, 122 (9), 465–468. Copyright 2002, with permission from Elsevier.

Interestingly, the observed BNNTs were capped with a W nanoparticle indicating a potential catalytic role in the growth process that was not understood at the time. HfB_2[282], YB_6[281], ZrB_2[283], and Ta[284] electrodes were also used to produce BNNTs using arc discharge and also reported metal particles acting as catalysts and being either encapsulated or at the tips of the tubes. Studies by Loiseau[282] and Terrones[284] primarily focused on the structure of the BNNTs themselves, and in particular, the various cap morphologies (including open-ended tube structures, a structural distinction from CNTs, which are always capped), to ascertain whether or not they contained topological defects similar to CNTs. These studies, as well as the work by Saito *et al.*,[285] observed a range of BNNT cap structures resulting from arc discharge synthesis including flag-shaped, bill-shaped, cone-shaped, and flat. These papers concluded that the caps must contain topological defects such as squares, pentagons, and heptagons to account for the disclinations observed *via* TEM (Fig. 3.28a,b). MD simulations performed by McLean *et al.*[164] present an alternative route to forming flat and angular BNNT caps, as discussed in Section 3.4.7.

More recently, arc discharge[286–288], and related plasma torch methods[289, 290] have looked to produce BNNTs and the related BN "nanococoons" (closed-shell BN capsules) on a macroscopic scale, revealing that a metal catalyst is not required for BNNT formation *via* this synthesis method.[288, 291] Synthesis of BNNTs using DC plasma has also revealed that H_2 plays an important role in mediating which growth species are present in the plasma, hence controlling BN growth.[292] MWBNNTs and SWBNNTs with diameters ranging from < 5 nm to 50 nm are observed, often with flat, angular caps (Fig. 3.28c).[287, 288] While experimental advances have been made, as have advances regarding our mechanistic understanding of BNNT growth *via* arc discharge, the production of large quantities of BNNTs using this method is hindered due to the extreme conditions required and the small growth volume. As a consequence, arc discharge synthesis of BNNTs has largely been superseded by alternative approaches.

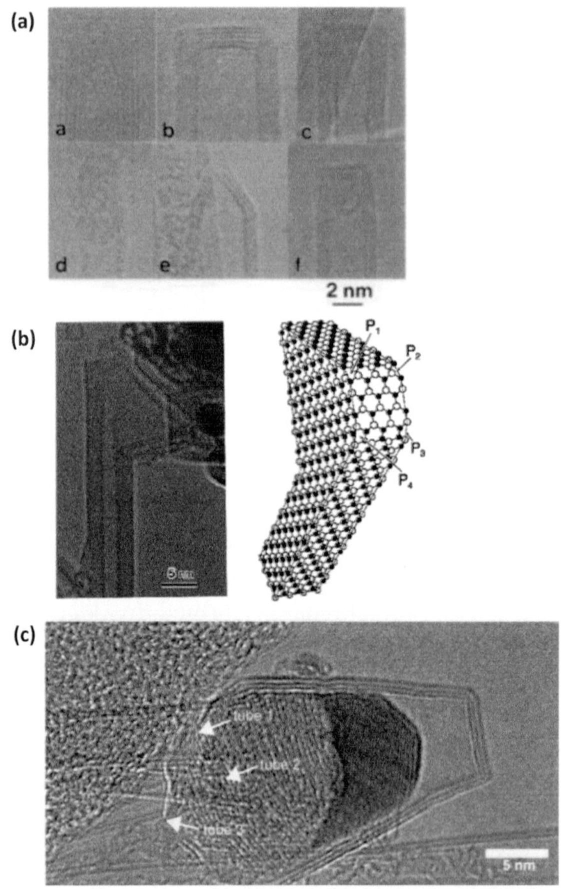

Figure 3.28 (a) TEM image of various BNNT cap morphologies resulting from arc discharge synthesis using HfB_2 electrodes in an N_2 atmosphere, featuring (i-iv, vi) rectangular, flat caps of varying diameter, and wall thickness, and (v) angular caps.[282] Reprinted with permission from Loiseau, A. *et al.* Boron nitride nanotubes with reduced numbers of layers synthesized by arc discharge. *Phys. Rev. Lett.* 1996, 76 (25), 4737–4740. Copyright 1996 by the American Physical Society. (b) Truncated "triangular flag"-shaped BNNT cap imaged by TEM following arc discharge synthesis using ZrB_2 electrodes in N_2 and the proposed structure, with four pentagons labeled P_1–P_4.[285] Reprinted with permission from Saito, Y. *et al.* Square, pentagon, and heptagon rings at BN nanotube tips. *J. Phys. Chem. A* 1999, 103 (10), 1291–1293. Copyright 1999 American Chemical Society. (c) STEM image of a BNNT attached to a B nanoparticle from the B-rich anode following arc discharge.[288] Distributed under a creative commons CC BY license.

3.4.2 Laser Ablation

Following arc discharge, laser heating methods were successful in the early growth of BNNTs. This technique was established in 1996 by Golberg and Bando, two names synonymous with the growth of BN nanomaterials, who reported MWBNNTs with outer cross-sections of 3–15 nm formed from using a laser to heat c-BN under GPa pressures.[293] In a similar fashion to arc discharge, the laser affords local synthesis temperatures to reach ~ 5,000 K, enabling gaseous precursors to combine in the gas phase to form BNNT growth species that deposit on the target. The laser ablation method, pioneered for the production of SWCNTs[294] was also used to produce MWBNNTs by ablating a powdered mixture of BN, Co, and Ni at 1,200 °C, which was then deposited onto a cooler Cu target downstream *via* a carrier gas.[295] While laser ablation is capable of producing SWBNNTs, relatively narrow-diameter tubes (1.5–8 nm), and high-density bundles, there are often impurities found in the final product, such as carbon, and there is little control over the number of SWBNNTs and MWBNNTs produced.[296–298]

While arc discharge produced the first SWBNNTs[282], early laser ablation studies using the Ni-Co catalyst also produced SWBNNTs, while those grown without a catalyst were multi-walled[299] and noticeably shorter in length.[300] Again, flatter caps were observed for BNNTs as well as encapsulated metal catalysts and a-BN clusters at the end of the tubes. The flatter cap structure of the BNNTs, compared to those of CNTs, was observed in numerous synthetic studies in the late 1990s and early 2000s and became a common trait of the material itself. While the morphology of the tubes was relatively constant, several studies showed that a catalyst was not required for the formation of SWBNNTs from CO_2 laser ablation of h-BN targets.[298, 301, 302] Lee *et al.* cited the presence of a pure B nanoparticle at the end of the as-formed SWBNNTs (Fig. 3.29a),[298] and Arenal *et al.*[297] proposed this as both the catalyst and the support for BNNT growth (Fig. 3.29b). In a detailed study of the growth mechanism, Arenal *et al.*[297] proposed that a liquid B nanoparticle would react with gaseous N_2 to nucleate the sp^2 BN cap on the surface, growing into a tube *via* a root-growth mechanism whereby N and B atoms are incorporated at the base of the BN network on the surface.[297] While this mechanism is accepted for closed BNNTs,[296] more recently Kim *et al.* proposed an open-end growth mechanism following their observations of uncapped BNNTs (Fig. 3.30a,b), whereby the B nanoparticle surface is first covered in h-BN *via* BN radical adsorption, and BN radicals are physisorbed at the exposed end of the tube.

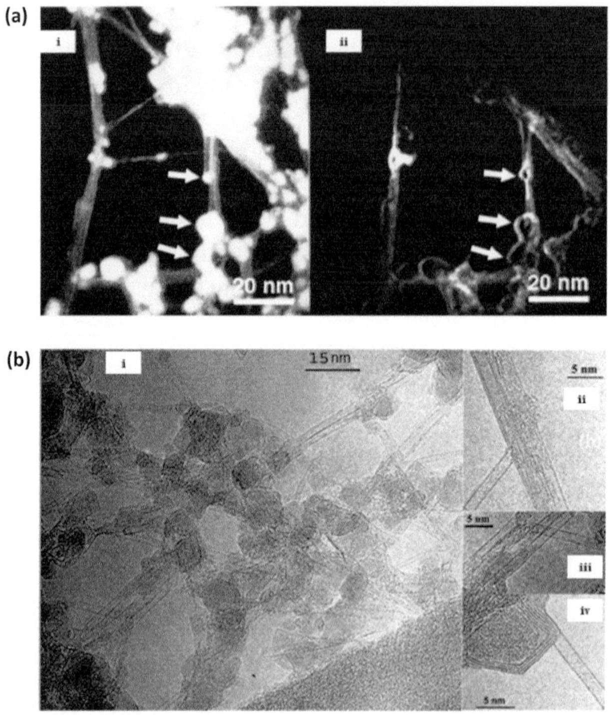

Figure 3.29 (a) EELs imaging of laser ablation-synthesized BNNTs providing a map of boron (A) and nitrogen (B), showing B nanoparticles encapsulated at the end of BNNTs or within a single BN layer shell (white arrows).[298] Reprinted with permission from Lee, R. S. *et al.* Catalyst-free synthesis of boron nitride single-wall nanotubes with a preferred zig-zag configuration. *Phys. Rev. B* 2001, 64 (12), 121405. Copyright 2001 by the American Physical Society. (b) TEM images of BNNT networks synthesized by Arenal. R. *et al.* show SWBNNTs, DWBNNTs, cages, and B nanoparticles acting as supports for these structures.[297] Reprinted with permission from Arenal, R. *et al.* Root-Growth Mechanism for single-walled boron nitride nanotubes in laser vaporization technique. *J. Am. Chem. Soc.* 2007, 129 (51), 16183–16189. Copyright 2007 American Chemical Society.

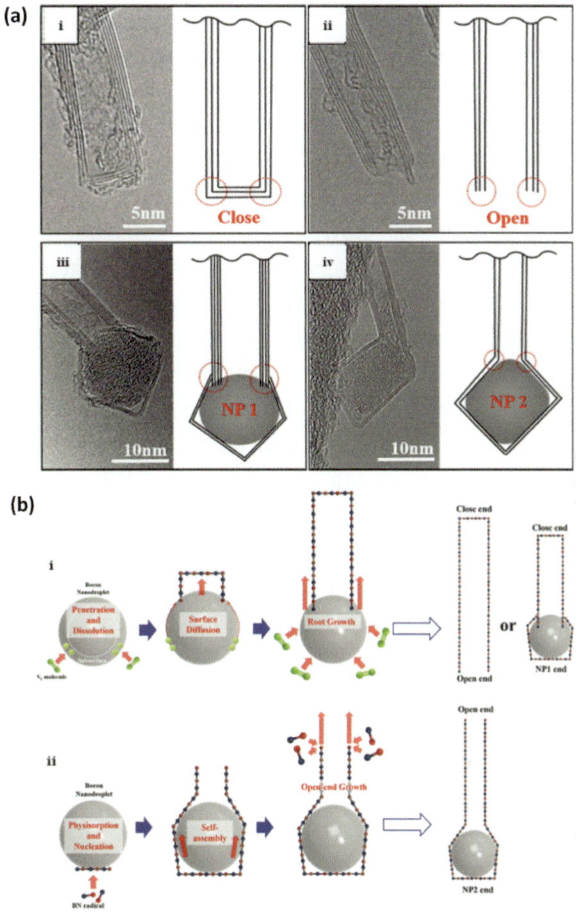

Figure 3.30 (a) TEM images of closed and open-ended BNNTs exhibiting two different morphologies with a coupled schematic showing the connectivity of the BNNT walls. (b) Root growth (top) and open-ended (bottom) growth mechanisms proposed by Kim *et al.* for laser ablation synthesis.[296] Distributed under a Creative Commons CC BY license.

3.4.3 Substitution Reactions

High-temperature methods such as arc discharge and laser ablation were predominant in the first decade of BNNT synthesis. However, their propensity to incorporate defects and impurities into the BNNT structures drove the investigation of alternative synthetic routes. One such route was chemical substitution based on CNTs, which attempted to exploit the fact that BNNTs are structurally analogous to CNTs. For instance, Han *et al.* reacted CNTs with vaporized B_2O_3 in an N_2 atmosphere at 1,773 K, converting all CNTs to BNNTs, with an equal amount of h-BN and r-BN formed as by-products.[303] The BNNTs thus-formed contained residual C, however, with subsequent methods improving to reach a conversion of 60%.[304] A more effective substitution was performed by Golberg *et al.*, who added a MoO_3 promoter and utilized lower temperatures.[305] While the MoO_3 helped to remove C impurities to an extent, it also resulted in Mo-encapsulated BNNTs. In an effort to steer away from C-based precursors, anodic aluminum oxide (AAO) supports saw success as templates for the growth of BNNTs and BN nanostructures from the pyrolysis of a 2,4,6-trichloroborzine precursor, which was injected into the pores of the AAO membrane.[274] AAO supports were also used as a template during the thermolysis of a borazine polymer and etched away to leave BNNTs and supported by a sheet of BN.[306] These processes produced highly ordered BN arrays, however, are subject to scale-up issues and often used harsh precursors and etching agents such as HF.[306]

3.4.4 Ball Milling

In an effort to increase the throughput and scale of BNNT synthesis, ball milling methods were developed by Chen *et al.*, forming BNNTs from precursors such as h-BN powder at 1,573 K,[307] and B powder in an atmosphere of NH_3 at 1,273 K.[308] The ball milling process utilizes stainless steel balls to grind the solid powder in quantities on the kilogram scale, inducing morphological and structural change. This process was often followed by annealing under N_2 at temperatures around 1,500 K. These processes saw yields of up to 85%[309] with the remaining product comprised polydisperse BN nanoparticles (Fig. 3.31a). The diameters of the tubes synthesized *via* ball milling vary from 20–200 nm and can exhibit bamboo-like or cone-shaped structures rather than typical BNNT cylinders.[310]

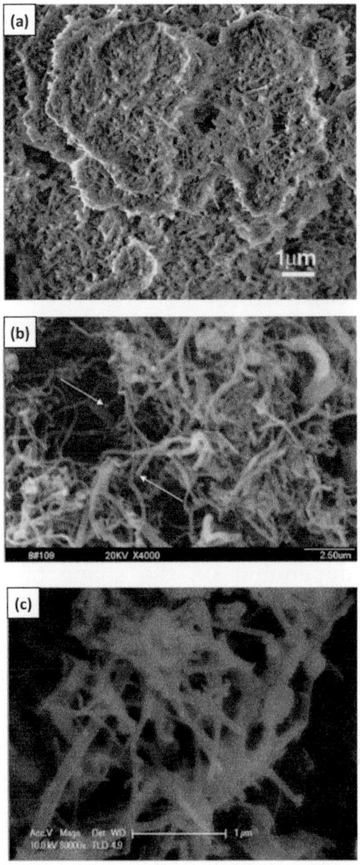

Figure 3.31 (a) SEM micrograph of BNNT and boron nanoparticle aggregates following ball milling in NH3.[309] Reprinted by permission from Springer Nature Customer Service Center GmbH: Springer Nature Chen, Y. *et al.* Large-quantity production of high-yield boron nitride nanotubes. *J. Mater. Res.* 2002, 17 (08), 1896–1899. Copyright 2002. (b) SEM image of BNNT networks following ball milling B2O3 in N2.[311] Reprinted from Li, Y. *et al.* Synthesis of boron nitride nanotubes from boron oxide by ball milling and annealing process. *Mater. Lett.* 2009, 63 (20), 1733–1736, Copyright 2009, with permission from Elsevier. (c) SEM image of typical BNNTs from ball milling B and N2.[312] Each SEM image shows the variation in BNNT morphology and diameter that comes with the ball milling process. Reprinted from Kim, J. *et al.* Synthesis and growth of boron nitride nanotubes by a ball milling-annealing process. *Acta Mater.* 2011, 59 (7), 2807–2813, Copyright 2011, with permission from Elsevier.

Ball milling can incorporate a metal catalyst, by combining it with the B precursor. For instance, Lim *et al.* milled a NiB_x powder with B to form a NiB_x catalyst prior to adding N_2 and H_2 gas at 1,500 K to form BNNTs.[313] The variation in BNNT diameter (20–40 nm) was reduced compared to prior studies, with H deemed crucial to the growth process. A multitude of catalysts (Al, Si, Fe, Ni, and Cr) assisted the formation of BNNTs from B_2O_3 and NH_3 at 1,500 K, though diameter variation was observed (Fig. 3.31b).[311]

The use of ball-milling is essentially a method of preparing the BNNT precursor, while the annealing process produces the BNNTs. Kim *et al.* showed that the post-annealing morphology of the BN can vary depending on the structure of the a-BN on the surface of the Fe catalyst (Fig. 3.31c).[312] The diameter of the BNNTs was dependent on the size of the seed particle, while the product morphology (i.e., nanotubes vs. nano-bamboo structures) was dependent on the shape of the seed particle. The lack of control over these two facets likely produced the variations in diameter and product structure observed in past studies. BNNT yield was also influenced by the starting structure produced from ball milling.[312] To avoid the limitation of ball milling producing variable precursor structures and products, the method has been combined successfully with arc discharge synthesis.[314]

3.4.5 Chemical Vapor Deposition

The most successful method for the synthesis of BNNTs is CVD. Typically performed at temperatures around 1,500 K or less, there is a greater degree of control compared with laser ablation and arc discharge methods, and as such, there has been greater interest in the CVD growth of BNNTs and the efforts at synthetic optimization. The first synthesis of BNNTs *via* CVD was reported in 2000 by Lourie *et al.*, using NiB catalyst particles and borazine ($B_3N_3H_6$) as the growth precursor (Fig. 3.32).[315] The borazine precursor was carried over the NiB particles on a Si substrate with N_2 gas at 1,373 K. The diameter of the BNNTs varied due to the polydispersity of the NiB particles. Though less successful, these authors also synthesized BNNTs using Co, Ni, and Ni_2B catalysts.[315] The BNNT structures were similar to that observed previously for arc discharge-synthesized tubes with sharper, flatter caps exhibiting flag-like shapes in some cases.[316] Ni[317] and Fe[318] borides also proved to be successful catalysts in later studies.

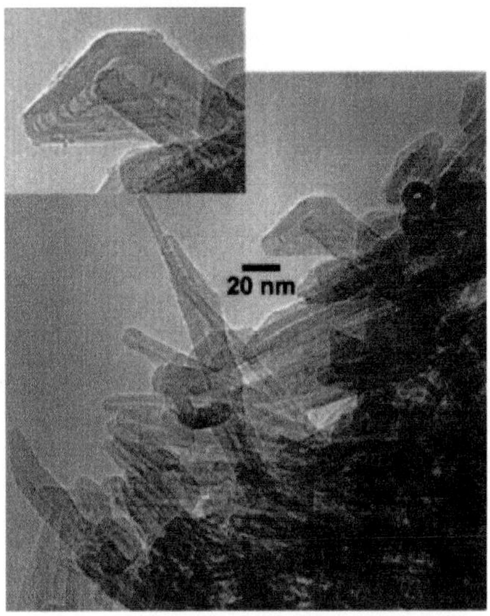

Figure 3.32 TEM image of the first CVD-grown BNNTs on NiB by Lourie O. R. *et al.*, showing flag-like and flat caps.[315] Reprinted with permission from Lourie, O. R. *et al.* CVD growth of boron nitride nanotubes. *Chem. Mater.* 2000, 12 (7), 1808–1810. Copyright 2000 American Chemical Society.

Early reports of BNNTs grown *via* CVD reported the use of oxides as growth promoters or catalysts, though afforded tubes with poor crystallinity or segmented structures, with little control over tube diameter. Growth promoters and catalysts used include V_2O_5,[319] Ag_2O,[320] MoO_3, CuO, PbO,[305, 321, 322] Fe_2O_3,[323] Fe-doped SiO_2,[324] and H_2O with stainless steel.[325] While not widely used, low-temperature microwave plasma-assisted CVD of B_2H_6/NH_3 achieved a degree of diameter control while maintaining near-ZZ alignment along the tube axis. Wang *et al.*[326] used Al_2O_3 substrates, Guo *et al.*[327, 328] employed Si(111) substrates coated with Co, Ni, and Fe, and Su *et al.*[329] employed Fe-supported SiO_2-Al_2O_3 catalysts. These methods produce

BNNTs at temperatures typically lower than CVD (< 1,273 K), however, struggle with respect to BNNT purity.

The utilization of a growth promoter in producing high-quality BNNTs was first realized in 2002 by Tang *et al.* *via* the boron oxide-assisted CVD method (BOCVD, Fig. 3.33a).[330] These growth promoters are typically metal oxide powders (MgO was used first,[330] although many followed, see list below) and react with solid B powder to form B_2O_2 *via* heating:

$$2MgO_{(s)} + 2B_{(s)} \rightarrow B_2O_{2(g)} + 2Mg_{(g)} \quad (3.1)$$

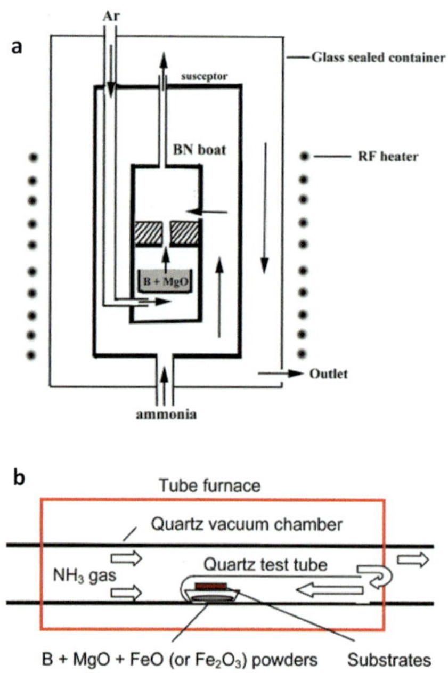

Figure 3.33 (a) Schematic of the RF induction furnace utilized by Tang *et al.* to first produce BNNTs *via* BOCVD. Arrows indicate directions of gas flow.[330] Tang, C.; Bando, Y.; Sato, T.; Kurashima, K., A novel precursor for synthesis of pure boron nitride nanotubes. *Chem. Commun.* 2002, 12 (12), 1290–1291. Published by The Royal Society of Chemistry. (b) Horizontal tube furnace used by Lee *et al.* to grow BNNTs *via* modified BOCVD. The growth vapor trapping technique was employed *via* the quartz tube having a closed-end in direction of incipient NH_3 gas flow.[342] Lee, C. H. *et al.* Effective growth of boron nitride nanotubes by thermal chemical vapor deposition. *Nanotechnology* 2008, 19 (45), 455605 © IOP Publishing. Reproduced with permission. All rights reserved.

The B_2O_2 gas reacts with ambient NH_3 in the reaction chamber to form BN on the walls of the reactor or on the surface of a catalyst according to:

$$B_2O_{2(g)} + 2NH_{3(g)} \rightarrow 2BN_{(s)} + 2H_2O_{(g)} + 2H_{2(g)} \qquad (3.2)$$

The BOCVD method initially afforded pure BNNTs at a yield of 40% with good control over tube diameter, with most exhibiting diameters of around 10 nm. The tubes were all MWBNNTs and were observed to have near-ZZ edges. The growth promoter, Mg, was assumed to have acted as the catalyst despite being absent from the end of the nanotubes.[330] The role of the oxide growth promoter sparked numerous synthetic routes to achieve BNNTs *via* BOCVD and saw the use of GaO,[331] FeO, Fe_2O_3,[332,333] V_2O_5/Fe_2O_3, V_2O_5/Ni_2O_3,[334] Al_2O_3, SnO,[335] $KFeO_2$,[336] $LiNO_3$,[337] and $(NH_4)_{10}W_{12}O_{41}{\cdot}xH_2O$.[338] To reduce the O content of the produced BNNTs and associated material, other growth promoters have since been trialed, such as Li_3N,[339] NH_4Cl, MgF_2,[340] and ZnS,[341] although the presence of S in the latter promoter led to SO_2 formation, which adversely modified the surface of the BNNTs.

The BOCVD method was subsequently improved by Lee *et al.* with the inclusion of a quartz tube within the reaction chamber, the chamber additionally had a closed end to the influx of precursor (Fig. 3.33b).[342] This traps the precursor vapors that enter the tube over the growth substrate and enhance BNNT formation. At 1,473 K, MWBNNTs were formed that were vertically aligned and highly crystalline (Fig. 3.34a).[342] With the addition of MgO, Ni, and Fe catalysts to the Si substrate, patterned growth of BNNTs was observed by the same group, highlighting the importance of catalytic action in this method (Fig. 3.34b).[343] Huang *et al.* employing the vapor trapping technique, replaced the MgO growth promoter with Li_2O and produced highly pure, crystalline MWBNNTs with diameters less than 10 nm, with lengths greater than 10 μm.[344] These studies were followed by a comprehensive experimental investigation by Pakdel *et al.* into the influence of precursor molar ratio (B:MgO:FeO) and growth temperature (1,473–1,673 K) on the morphology, diameter, composition, crystallinity, and cap structure of BOCVD-synthesized BNNTs.[345] Lower growth temperatures here produced straighter and narrower BNNTs and the diameter could be increased through an increase in catalyst content. Again, an array of BNNT cap structures were observed, albeit flatter and more angular than those for CNTs, consistent with the BNNT morphologies obtained using alternative synthesis methods such as arc discharge (Section 3.4.1). Figure 3.35 shows the influence of metal oxide growth promoter: B ratio and growth temperature on the morphology of the films, the outer diameter of the BNNT walls, and the cap structure demonstrated by Pakdel *et al.*[345]

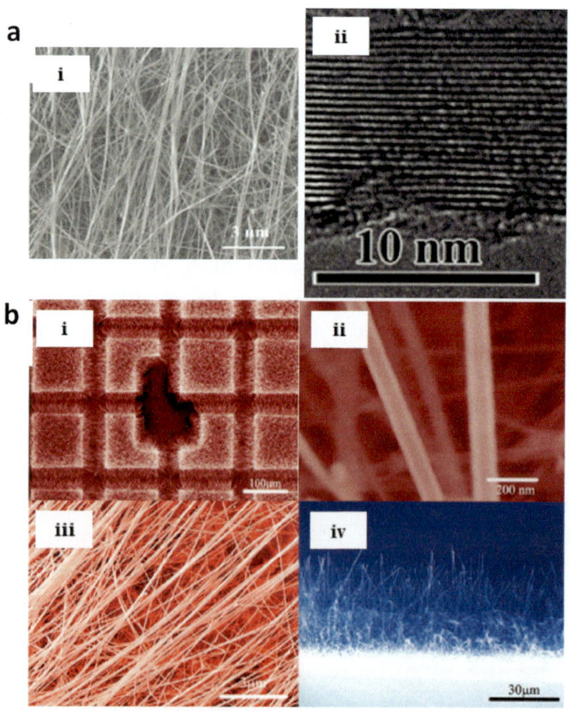

Figure 3.34 (a) SEM images of BNNTs grown in an alumina boat by Lee *et al.* using BOCVD.[342] Lee, C. H. *et al.* Effective growth of boron nitride nanotubes by thermal chemical vapor deposition. *Nanotechnology* 2008, 19 (45), 455605 © IOP Publishing. Reproduced with permission. All rights reserved. (b) SEM images (artificial color) of patterned growth of BNNTs on MgO, Ni, and Fe by Lee *et al.* using BOCVD.[343] Reprinted with permission from Lee, C. H. *et al.* Patterned growth of boron nitride nanotubes by catalytic chemical vapor deposition. *Chem. Mater.* 2010, 22 (5), 1782–1787. Copyright 2010 American Chemical Society.

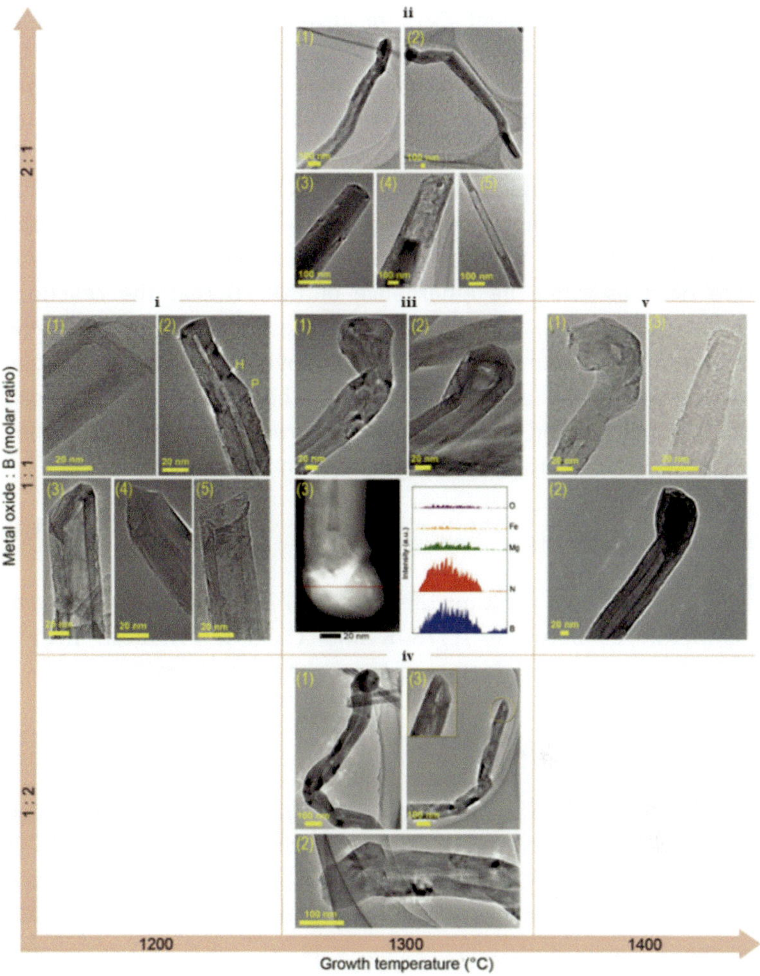

Figure 3.35 TEM analysis by Pakdel *et al.*[345] of the morphology of the BNNT cap as a function of experimental parameters. (i) Most commonly observed, flat BNNT cap. (ii) Long and wide BNNTs containing multiple kinks grown at a 2:1 metal oxide: B ratio. (iii) Caps encapsulating nanoparticles with elemental analysis showing catalyst and growth promoter content. (iv) BNNTs containing multiple kinks and curves. (v) Diameter-related tip morphology with flat caps in the absence of nanoparticles. Pakdel, A. *et al.* A comprehensive analysis of the CVD growth of boron nitride nanotubes. *Nanotechnology* 2012, 23 (21), 215601–215610

Although the use of CVD for BNNT growth had evolved in the decade following its first use, there was still disagreement regarding the mechanism by which BNNTs were forming. It was unclear whether the metal catalysts that allowed for patterned growth[343] or the growth promoters utilized[330, 344, 345] were controlling the formation of BNNTs. The most commonly used growth promoter was MgO, and while it was proposed that MgB_2 or MgN_2 may form during the process,[343] it was almost two decades after the emergence of BOCVD that the reactions involving the growth promoter (i.e., Eqs. 3.1 and 3.2) became a focus. The group of Yao *et al.* employed an MgB_2 catalyst to more successfully grow BNNTs than when using MgO or MgN_2 at 1,573 K,[338] and subsequently lowered operational costs by generating MgB_2 *in situ* with Mg rather than MgO:[338, 346]

$$4Mg_{(g)} + B_2O_{3(g)} \rightarrow MgB_{2(g)} + 3MgO_{(s)}. \tag{3.3}$$

Subsequently reacting the promoter with B_2O_3 and NH_3 at temperatures between 1,373 K and 1,673 K produced BNNTs of high quality and diameters of approximately 40 nm:[346]

$$B_2O_{3(g)} + 2NH_{3(g)} \rightarrow 2BN_{(s)} + 3H_2O_{(g)}. \tag{3.4}$$

The key role of the growth promoter as a catalyst is revealed here, without MgB_2, h-BN was synthesized rather than BNNTs. The generation of the catalyst *in situ* presents a unique and promising route for BNNT synthesis, as floating catalyst CVD using a generated nickelocene catalyst with borazine and decaborane has successfully formed double-walled BNNTs with attractive properties, though the method has not been thoroughly investigated.[347, 348]

As discussed, most BNNT growth strategies require some manner of catalyst to be successful, most often a metal. Alternative approaches have been taken to synthesize BNNTs *via* CVD including the use of a $C_3N_6H_6{\cdot}2H_3BO_3$ precursor by Ma *et al.*,[349] where amorphous BNO clusters formed and acted as supports and catalysts for growth. As discussed in Section 3.4.3, substitution reactions and the use of templates can also produce BNNTs. CNTs were used as a template for the CVD growth of

MWBNNTs from boric acid and NH_3.[350] Both methods, however, resulted in the contamination of the product with C and O, a phenomenon rarely observed when a metal catalyst is used.

3.4.6 BNNT Growth Mechanism

Since the first synthesis of BNNTs, there have been many advances in each synthetic technique, particularly CVD, and the understanding of how experimental parameters such as temperature, precursors, catalysts, and growth promoters influence the final product. However, the mechanism by which a BNNT nucleates and grows is, by comparison, poorly understood. Experimental studies discussed thus far have provided insight and have revealed broad mechanistic trends, though it is when this work is combined with theoretical insight from computational chemistry that the understanding is truly enhanced, as was the case for the field of CNT growth.[351]

Generally, BNNTs can catalytically grow in one of two ways as described in the literature: "tip-growth", where the BNNT grows between the catalyst and the substrate, often with the catalyst nanoparticle observed encapsulated in the cap of the BNNT; or "root-growth", where the catalyst-substrate interaction is strong and BNNT grows outwards, with the catalyst remaining at the base. Following the first observation of BNNTs from CVD, Lourie *et al.*[315] proposed a root-growth mechanism, where the BNNT nucleated on the surface of the NiB particles and grew outwards, as they observed the angular, flag-like caps that contained so metal particles (Fig. 3.36). It was proposed that these cap structures grew initially on the catalyst and were preserved as the BNNT grew outwards, i.e., extruded away, from the surface. Conversely, more bulb-like caps were observed following the aforementioned catalyst-free growth by Ma *et al.*[349] and as such a tip-growth mechanism was proposed. Here, the a-BN formed from the precursor, which itself acted as a catalyst but was lifted from the substrate along with the tube.

Following the use of the BOCVD method, with an explicit growth promoter and/or catalyst was used, the vapor-liquid-solid (VLS) mechanism[352] was widely accepted, regardless of the tip- or root-growth mode. For example, this agreed with the observed root-growth mechanism by Lourie *et al.*,[315] where the precursors dissolve into the molten NiB catalyst before reacting with N-containing precursors to form BN. However, the study by Pakdel *et al.*[345] exploring the parameter space and the effect on the final product proposed a VLS mechanism with a regime dependent on the parameters for growth, primarily temperature (Fig. 3.37). A low growth temperature (1,473 K) saw a root-growth mechanism dominate as BN saturated the catalyst surface before precipitating into BN shells and tubes, layer by layer. With an increase in temperature (1,573 K), the catalyst particles are observed in the caps of the BNNTs off the surface, hence a tip-growth mechanism was adopted. Finally, increasing the temperature to 1,673 K saw multiple nucleation sites become apparent on the same catalyst particle as multiple BNNTs would be formed with varying structures, indicating a mixed growth regime. At this temperature, the size of the catalyst particles would increase due to the enhanced aggregation of metal atoms on the substrate *via* Ostwald ripening.

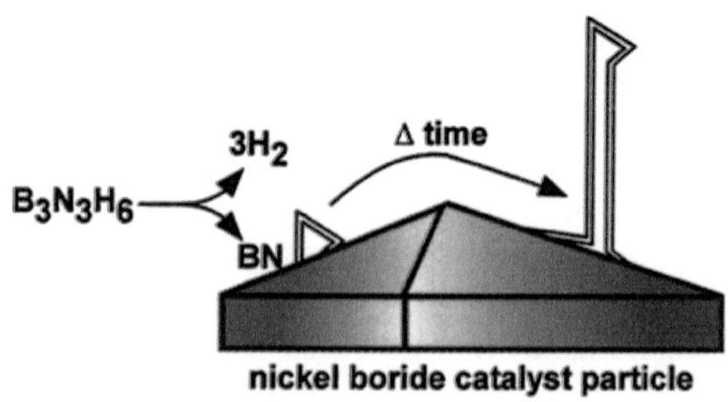

Figure 3.36 Root-growth mechanism of BNNT growth proposed by Lourie *et al.*, with incomplete BN layers at the nanotube-catalyst interface.[315] Reprinted with permission from Lourie, O. R. *et al.* CVD growth of boron nitride nanotubes. *Chem. Mater.* 2000, 12 (7), 1808–1810. Copyright 2000 American Chemical Society.

A solid-liquid-solid (SLS) mechanism has instead been proposed by Lahiri *et al.*[353] for the Cu-catalyzed growth of BNNTs from boron oxide and nitrogen-containing species (most often NH_3), though the mechanism of O release from the pure BNNTs was not specified. It is generally accepted that the growth of BNNTs is *via* a VLS mechanism.[351]

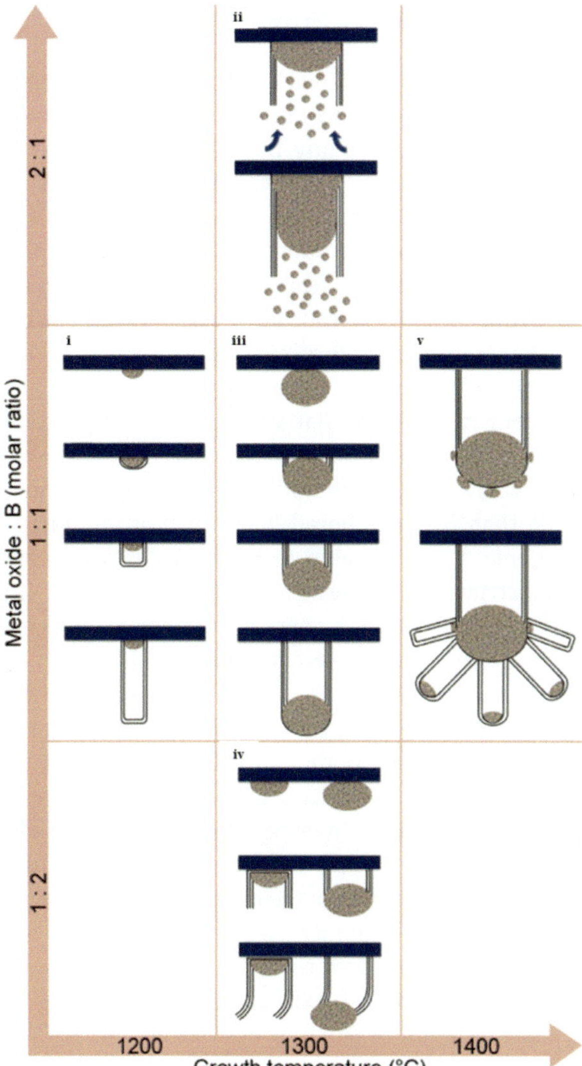

Figure 3.37 Growth mechanisms suggested by Pakdel, A. *et al.* for BNNTs as a function of experimental parameters. (i) Root-growth mechanism. (ii) Tube filling during a root-growth model. (iii) Tip-growth model. (iv) Changes in the growth direction during growth by either root- or tip-growth. (v) Mixed-growth model, with multiple nucleation sites.[345] Pakdel, A. *et al.* A comprehensive analysis of the CVD growth of boron nitride nanotubes. *Nanotechnology* 2012, 23 (21), 215601–215610

3.4.7 Theoretical BNNT Growth Mechanism

Prior to studying the growth mechanism of BNNTs during growth, theoretical efforts focused on understanding their structure and energetics. The first such study by Blase *et al.*[354] used *ab initio* MD simulations to investigate the stability that (5,5) armchair BNNT cap structures containing even-numbered BN rings at high temperatures. Upon addition of B, N, and BN groups to the caps, hexagonal BN rings were formed *via* an intermediate pentagon. From a (9,0) ZZ cap, however, 4-membered BN rings resulted. The ZZ cap, therefore, was concluded to be less stable and likely to self-assemble during growth into an amorphous cap, in agreement with the flat caps observed by Loiseau *et al.* following laser ablation[282] and arc discharge.[316] Even-membered rings, specifically 4|8|-membered ring pairs, were shown to be more stable than 5|7-membered pairs even at low temperatures, and undergoing a process akin to Stone-Wales transformation in carbon nanostructures to form a purely hexagonal BNNT above 1,500 K.[355] Several studies agreed that with odd-membered rings, the presence of homoelemental (B-B, N-N) bonds frustrated the closure of cap closures and led to overall instability. Of the even-membered rings, 4-membered were shown to be the least stable, leading to the formation of open H-passivated edges rather than a closed cap, to avoid geometrical strain.[356] Figure 3.38 shows the investigated formation and growth mechanism scheme of a (3,3) BNNT from a $B_{12}N_{12}$ fullerene by Xu *et al.*[356] using DFT, though in the absence of a catalyst.

A thermochemical model was proposed by Kuznetsov *et al.* following DFT calculations involving three proposed BNNT cap nuclei.[357] These nuclei explored three possible centers of the cap: a BN hexagon, a single B or N atom, and a BN octagon. Placing hexagons around these centers in different directions allowed both armchair and ZZ tubes to be achieved. The model highlighted reaction temperature, degree of catalyst particle supersaturation, the adhesion strength between the metal and the h-BN phase, and the B/N-metal binding energies as critical BNNT growth parameters. While this model focused on even-membered rings in most cases, DFT calculations from Wu *et al.*[358] and Zhao *et al.*[359] concluded independently that the exothermic chemisorption of BNNTs on metal surfaces (e.g., Ni) was facilitated by defects within the tubes. McLean *et al.*[360] investigated the adsorption of the armchair and ZZ BNNTs on 1st, 2nd, and 3rd row transition metal catalyst nanoparticles and revealed charge transfer at the interface to be the key determinant of the strength of the BNNT-catalyst interaction. The ZZ BNNT catalyst was revealed to be more stable than the armchair edge while the preference for the N-terminated or B-terminated edge can be tuned with the identity of the catalyst (Fig. 3.39). Mg has also proved an effective catalyst for BNNT growth during BOCVD (discussed above) and E *et al.*[338] used DFT and *ab initio* calculations to show that MgB_2 formed from the reaction between Mg and B dissolves B_xO_x to afford BO and BO_2 that would subsequently react on the surface with NH_3 to afford BN (Fig. 3.40d).

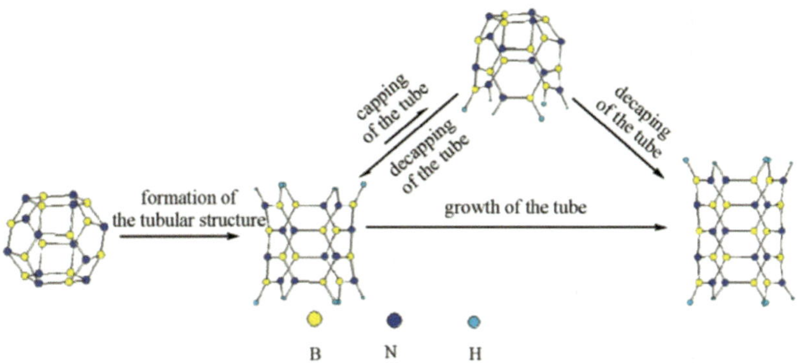

Figure 3.38 Formation and growth of a (3,3) BNNT from the opening of a $B_{12}N_{12}$ fullerene and addition of either B_2H_6, NH_3, or H_2.[356] Reprinted with permission from Xu, H. *et al.* The electronic structures and formation mechanisms of the single-walled BN nanotube with small diameter. *J. Phys. Chem. B* 2004, 108 (13), 4024–4034. Copyright 2004 American Chemical Society.

Regarding the mechanisms of nucleation and growth, gas-phase interactions (in the absence of a catalyst) were investigated by Weismiller *et al.*[361] using reactive MD simulations (Fig. 3.40a). These simulations examined conditions that span multiple synthetic techniques (300–5,000 K) and demonstrated how a common BN precursor, ammonia borane, decomposes to release H_2, and polymerises at high temperatures. This work also demonstrated the clustering of B and O in the gas-phase over the same range of temperatures, primarily resulting in B-based clusters at CVD temperatures (700–1,600 K) amorphous BO-based clusters at high temperatures (> 2,000 K). Focusing on CVD growth temperatures, McLean *et al.*[362] employed non-equilibrium reactive MD simulations to investigate the initial reactive pathways during BOCVD (Fig. 3.40c). This work demonstrated how the B_2O_2 and NH_3 growth precursors nucleate BN in the gas phase as well the key roles of H_2 and H_2O in the process. The release of H_2 affords the clustering of B-rich clusters, which can support the growth of $B_xN_yO_xH_z$ chains, however, the strength of the B-O bond prevents it from being cleaved without the influence of a catalyst or the manual removal of H_2O from the simulations, which drove the formation of B-N bonds and a-BN networks.

High-temperature methods were simulated by Krstic *et al.*[363] using non-equilibrium MD (Fig. 3.40b). At temperatures greater than 2,000 K, atomic BN, BNH-based precursors including borazine self-assemble into BN sheets, cages, and tube-like structures with defects healing into hexagons. In the absence of a catalyst for BN nanostructures, the "pentagon-first" nucleation mechanism observed for carbon nanostructures[364] did not hold. At high temperatures, H appeared to not greatly influence the formation mechanism, though it was previously suggested to catalyze the conversion between c-BN and h-BN.[365] The addition of N atoms to a B slab (mimicking a B nanoparticle) in high-temperature (> 2,000 K) *ab initio* MD simulations by Santra *et al.*[291] showed a root-growth mechanism for the formation of a BN island on the surface and the subsequent growth of a BNNT as B and N atoms were incorporated into the base. Interestingly, only even-membered rings were observed during these plasma-temperature simulations.

Ni-catalyzed CVD BNNT growth mechanisms were investigated by McLean *et al.* for ammonia borane CVD[164] and BOCVD[366] using non-equilibrium reactive MD simulations. The catalytic action of Ni was required to release H_2 from the ammonia borane precursor and facilitate the formation of defect-free BN ring networks *via* the cleavage of homoelemental B-B and N-N bonds.[164] For BOCVD, the presence of surface-bound O and the B-O bond strength hindered the formation of extended networks without the removal of H_2O.[366] However, in both BOCVD and ammonia borane CVD, once BN ring networks are formed, BNNT cap structures nucleate *via* a network fusion mechanism such that the ZZ edges of the BN sheets fuse together as they are anchored perpendicular to the surface (Fig. 3.41d). Importantly, these cap structures grow *via* surface diffusion and contain only hexagons, without 4|8 pairs or pentagon defects at the cap corners.

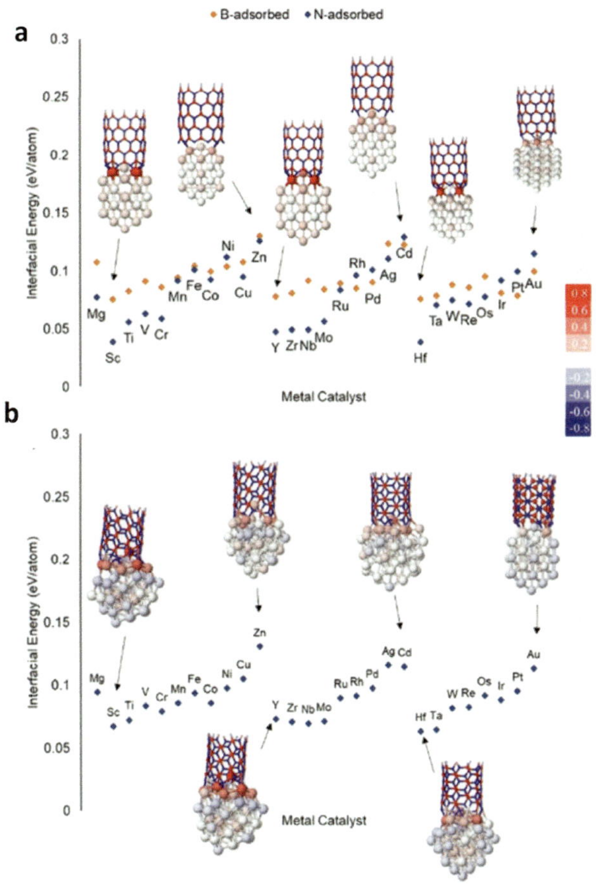

Figure 3.39 Interfacial energies (eV/edge atom) as a function of catalysts identity for (a) N- and B-adsorbed (10,0) BNNTs and (b) (5,5) BNNTs on 55 atom metal nanoparticle catalysts. Atom colors per scale bar according to their net charge.[360] Reprinted with permission from McLean, B. *et al.* Energy and charge transfer at the boron nitride nanotube—catalyst growth interface. *J. Phys. Chem. C* 2020, 124 (21), 11662–11668. Copyright 2020 American Chemical Society.

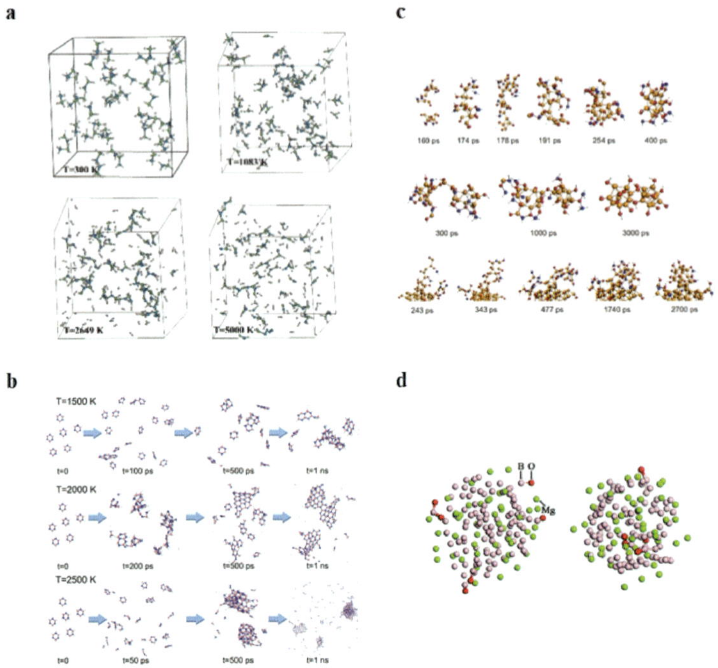

Figure 3.40 (a) Temperature ramp reactive MD simulations by Weismiller, M. R. *et al.* observing the decomposition of ammonia borane around CVD growth temperatures and polymerization at temperatures > 2,000 K.[361] Reprinted with permission from Weismiller, M. R. *et al.* ReaxFF reactive force field development and applications for MD simulations of ammonia borane dehydrogenation and combustion. *J. Phys. Chem.* 2010, 114 (17), 5485–5492. Copyright 2010 American Chemical Society. (b) High-temperature MD simulations by Krstic, P. S. *et al.* observing the nucleation of BN ring networks from a borazine precursor at 1,500 K and 2,000 K and the formation of curved structures at 2,500 K.[363] Krstic, P. S. *et al.* Simulations of the synthesis of boron-nitride nanostructures in a hot, high-pressure gas volume. *Chem. Sci.* 2018, 9 (15), 3803–3819. Published by The Royal Society of Chemistry. (c) Reactive MD simulations by McLean, B. *et al.* revealing B nanoparticle formation during BOCVD *via* B_2O_2 and NH_3 reaction at 1,373 K.[362] Reprinted with permission from McLean, B. *et al.* Boron nitride nucleation mechanism during chemical vapor deposition. *J. Phys. Chem. C* 2018, 122 (42), 24341–24349. Copyright 2018 American Chemical Society. (d) *Ab initio* MD simulations by E. S. *et al.* showing the dissolution of BO groups into an MgB_2 nanoparticle.[338] E. S. *et al.* Growth of boron nitride nanotubes from magnesium diboride catalysts. *Nanoscale* 2018, 10 (29), 13895–13901. Published by The Royal Society of Chemistry.

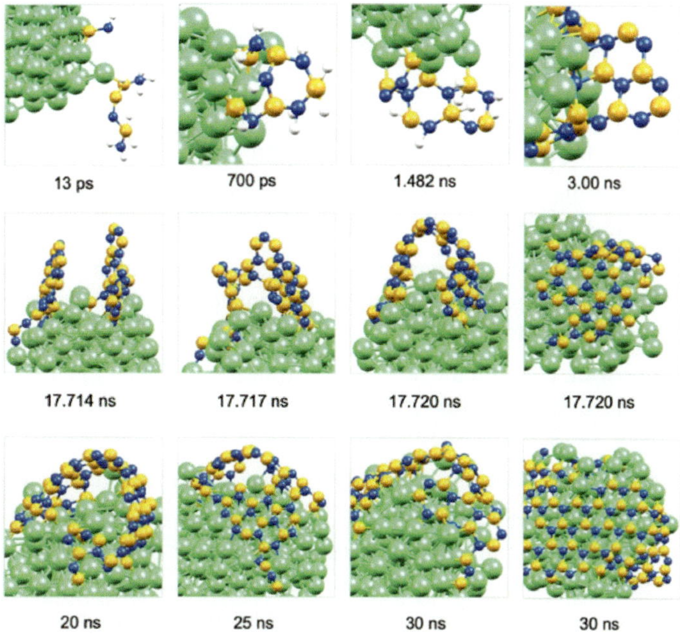

Figure 3.41 Structural evolution of a BN ring network from a BN chain on a Ni catalyst observed by McLean, B. *et al. via* reactive MD simulations and reveal of the BNNT cap formation mechanism *via* network fusion, in the absence of topological defects such as BN pentagons and heptagons.[164] Reprinted with permission from McLean, B. *et al.* Boron nitride nanotube nucleation *via* network fusion during catalytic chemical vapor deposition. *J. Am. Chem. Soc.* 2019, 141 (34), 13385–13393. Copyright 2019 American Chemical Society.

References

1. Wentorf, R. H., Cubic form of boron nitride. *J. Chem. Phys.* **1957**, *26* (4), 956–956.
2. Wentorf, R. H., Synthesis of the cubic form of boron nitride. *J. Chem. Phys.* **1961**, *34* (3), 809–812.
3. Ulrich, S.; Ye, J.; Stüber, M.; Ziebert, C., Cubic boron nitride based metastable coatings and nanocomposites. *Thin Solid Films.* **2009**, *518* (5), 1443–1450.
4. Vel, L.; Demazeau, G.; Etourneau, J., Cubic boron nitride: Synthesis, physicochemical properties and applications. *Mater. Sci. Eng., B.* **1991**, *10* (2), 149–164.

5. Mirkarimi, P. B.; McCarty, K. F.; Medlin, D. L., Review of advances in cubic boron nitride film synthesis. *Mater. Sci. Eng. R.* **1997**, *21* (2), 47–100.
6. Zhang, W. J.; Bello, I.; Lifshitz, Y.; Lee, S. T., Recent advances in cubic boron nitride deposition. *MRS Bull.* **2003**, *28* (3), 184–188.
7. Wakatsuki, M.; Ichinose, K.; Aoki, T., Synthesis of polycrystalline cubic BN. *Mater. Res. Bull.* **1972**, *7* (9), 999–1003.
8. Wakatsuki, M.; Ichinose, K., Synthesis of polycrystalline cubic BN (vi)–some features of transformations to cub BN and to Wur BN. **1975**, 441–445.
9. Corrigan, F. R.; Bundy, F. P., Direct transitions among the allotropic forms of boron nitride at high pressures and temperatures. *J. Chem. Phys.* **1975**, *63* (9), 3812–3820.
10. Sirota, N. N., Process for production of cubic boron nitride. 1973.
11. Sumiya, H.; Iseki, T.; Onodera, A., High pressure synthesis of cubic boron nitride from amorphous state. *Mater. Res. Bull.* **1983**, *18* (10), 1203–1207.
12. Gladkaya, I. S.; Kremkova, G. N.; Slesarev, V. N., Turbostratic boron nitride (BNt) under high pressures and temperatures. *J. Less. Comm. Met.* **1986**, *117* (1), 241–245.
13. Dubrovinskaia, N.; Solozhenko, V. L.; Miyajima, N.; Dmitriev, V.; Kurakevych, O. O.; Dubrovinsky, L., Superhard nanocomposite of dense polymorphs of boron nitride: Noncarbon material has reached diamond hardness. *Appl. Phys. Lett.* **2007**, *90* (10), 101912.
14. Tian, Y.; Xu, B.; Yu, D.; Ma, Y.; Wang, Y.; Jiang, Y.; Hu, W.; Tang, C.; Gao, Y.; Luo, K.; Zhao, Z.; Wang, L.-M.; Wen, B.; He, J.; Liu, Z., Ultrahard nanotwinned cubic boron nitride. *Nature.* **2013**, *493* (7432), 385–388.
15. Mirkarimi, P. B.; McCarty, K. F.; Medlin, D. L.; Wolfer, W. G.; Friedmann, T. A.; Klaus, E. J.; Cardinale, G. F.; Howitt, D. G., On the role of ions in the formation of cubic boron nitride films by ion-assisted deposition. *J. Mater. Res.* **1994**, *9* (11), 2925–2938.
16. Medlin, D. L.; Friedmann, T. A.; Mirkarimi, P. B.; Rez, P.; Mills, M. J.; McCarty, K. F., Microstructure of cubic boron nitride thin films grown by ion-assisted pulsed laser deposition. *J. Appl. Phys.* **1994**, *76* (1), 295–303.
17. Mineta, S.; Kohata, M.; Yasunaga, N.; Kikuta, Y., Preparation of cubic boron nitride film by CO_2 laser physical vapour deposition with

simultaneous nitrogen ion supply. *Thin Solid Films.* **1990**, *189* (1), 125–138.

18. Friedmann, T. A.; Mirkarimi, P. B.; Medlin, D. L.; McCarty, K. F.; Klaus, E. J.; Boehme, D. R.; Johnsen, H. A.; Mills, M. J.; Ottesen, D. K.; Barbour, J. C., Ion-assisted pulsed laser deposition of cubic boron nitride films. *J. Appl. Phys.* **1994**, *76* (5), 3088–3101.

19. Ballal, A. K.; Salamanca-Riba, L.; Doll, G. L.; Taylor, C. A.; Clarke, R., Ion-assisted pulsed laser deposition of cubic BN films on Si (001) substrates. *J. Mater. Res.* **1992**, *7* (7), 1618–1620.

20. Narayan, J.; Bhaumik, A.; Xu, W., Direct conversion of h-BN into c-BN and formation of epitaxial c-BN/diamond heterostructures. *J. Appl. Phys.* **2016**, *119* (18), 185302.

21. Narayan, J.; Bhaumik, A., Research update: Direct conversion of h-BN into pure c-BN at ambient temperatures and pressures in air. *APL Mater.* **2016**, *4* (2), 020701.

22. Haque, A.; Narayan, J., Conversion of h-BN into c-BN for tuning optoelectronic properties. *Mater. Adv.* **2020**, *1* (4), 830–836.

23. Mamoru, S.; Fuminori, F., Formation of cubic boron nitride films by boron evaporation and nitrogen ion beam bombardment. *Jpn. J. Appl. Phys.* **1983**, *22* (3A), L171.

24. Erler, H. J.; Reisse, G.; Weissmantel, C., Nitride film deposition by reactive ion beam sputtering. *Thin Solid Films.* **1980**, *65* (2), 233–245.

25. Inagawa, K.; Watanabe, K.; Ohsone, H.; Saitoh, K.; Itoh, A., Preparation of cubic boron nitride film by activated reactive evaporation with a gas activation nozzle. *J. Vac. Sci. Technol. A.* **1987**, *5* (4), 2696–2700.

26. Yokoyama, H.; Okamoto, M.; Osaka, Y., Effects of a negative self-bias on the growth of cubic boron nitride prepared by plasma chemical vapor deposition. *Jpn. J. Appl. Phys.* **1991**, *30* (2R), 344.

27. Kester, D. J.; Messier, R., Phase control of cubic boron nitride thin films. *J. Appl. Phys.* **1992**, *72* (2), 504–513.

28. Kester, D. J.; Ailey, K. S.; Davis, R. F.; More, K. L., Phase evolution in boron nitride thin films. *J. Mater. Res.* **1993**, *8* (6), 1213–1216.

29. Kester, D. J.; Ailey, K. S.; Lichtenwalner, D. J.; Davis, R. F., Growth and characterization of cubic boron nitride thin films. *J. Vac. Sci. Technol. A.* **1994**, *12* (6), 3074–3081.

30. Kester, D. J.; Ailey, K. S.; Davis, R. F., Deposition and characterization of boron nitride thin films. *Diamond Relat. Mater.* **1994**, *3* (4), 332–336.

31. Tanabe, N.; Hayashi, T.; Iwaki, M., Substrate temperature influence of c-BN thin film formation by IBED. *Diamond Relat. Mater.* **1992**, *1* (2), 151–156.

32. Ikeda, T., Cubic boron nitride films synthesized by low-energy ion-beam-enhanced vapor deposition. *Appl. Phys. Lett.* **1992**, *61* (7), 786–788.

33. Reinke, S.; Kuhr, M.; Kulisch, W., Mechanisms in ion induced c-BN growth. *Diamond Relat. Mater.* **1994**, *3* (4), 341–345.

34. Reinke, S.; Kuhr, M.; Kulisch, W., Critical test of the c-BN sputter model. *Surf. Coat. Technol.* **1995**, *74–75*, 723–728.

35. Reinke, S.; Kuhr, M.; Kulisch, W.; Kassing, R., Recent results in cubic boron nitride deposition in light of the sputter model. *Diamond Relat. Mater.* **1995**, *4* (4), 272–283.

36. McKenzie, D. R.; McFall, W. D.; Smith, H.; Higgins, B.; Boswell, R. W.; Durandet, A.; James, B. W.; Falconer, I. S., High pressure phases produced by low energy ion implantation with reference to cubic boron nitride. *Nucl. Instrum. Meth. B.* **1995**, *106* (1), 90–95.

37. Hahn, J.; Richter, F.; Pintaske, R.; Röder, M.; Schneider, E.; Welzel, T., Formation of c-BN thin films under reduced ion impact. *Surf. Coat. Technol.* **1997**, *92* (1), 129–134.

38. Djouadi, M. A.; Bouchier, D.; Moller, P.; Sene, G., Antibes. In *9th Int. Colloq. on Plasma Processes*. 1993.

39. Mieno, M.; Yoshida, T., Preparation of cubic boron nitride films by RF sputtering. *Jpn. J. Appl. Phys.* **1990**, *29* (7A), L1175.

40. Kulikovsky, V. Y.; Shaginyan, L. R.; Vereschaka, V. M.; Hatynenko, N. G., Preparation of thin hard boron nitride films by RF magnetron sputtering. *Diamond Relat. Mater.* **1995**, *4* (2), 113–119.

41. Schütze, A.; Bewilogua, K.; Lüthje, H.; Kouptsidis, S.; Jäger, S., Cubic boron nitride films prepared by reactive RF and DC sputtering from different boron containing targets. *Surf. Coat. Technol.* **1995**, *74–75*, 717–722.

42. Tsuda, O.; Yamada, Y.; Fujii, T.; Yoshida, T., Preparation of cubic boron nitride films by radio frequency bias sputtering. *J. Vac. Sci. Technol. A.* **1995**, *13* (6), 2843–2847.

43. Ulrich, S.; Scherer, J.; Schwan, J.; Barzen, I.; Jung, K.; Scheib, M.; Ehrhardt, H., Preparation of cubic boron nitride films by radio frequency magnetron sputtering and radio frequency ion plating. *Appl. Phys. Lett.* **1996**, *68* (7), 909–911.

44. Gimeno, S.; Andújar, J. L.; Lousa, A., Cubic boron nitride thin films by tuned RF magnetron sputtering. *Diamond Relat. Mater.* **1997**, *6* (5), 604–607.

45. Mirkarimi, P. B.; Medlin, D. L.; McCarty, K. F.; Dibble, D. C.; Clift, W. M.; Knapp, J. A.; Barbour, J. C., The synthesis, characterization, and mechanical properties of thick, ultrahard cubic boron nitride films deposited by ion-assisted sputtering. *J. Appl. Phys.* **1997**, *82* (4), 1617–1625.

46. Yamada, Y.; Tatebayashi, Y.; Tsuda, O.; Yoshida, T., Growth process of cubic boron nitride films in bias sputter deposition. *Thin Solid Films.* **1997**, *295* (1), 137–141.

47. Ye, J.; Rothhaar, U.; Oechsner, H., Conditions for the formation of cubic boron nitride films by RF magnetron sputtering. *Surf. Coat. Technol.* **1998**, *105* (1), 159–164.

48. Tzeng, Y.; Zhu, H., Electron-assisted deposition of cubic boron nitride by RF magnetron sputtering. *Diamond Relat. Mater.* **1999**, *8* (8), 1402–1405.

49. Deng, J.; Wang, B.; Tan, L.; Yan, H.; Chen, G., The growth of cubic boron nitride films by RF reactive sputter. *Thin Solid Films.* **2000**, *368* (2), 312–314.

50. Jiang, L.; Fitzgerald, A. G.; Rose, M. J.; Lousa, A.; Gimeno, S., Formation of cubic boron nitride films by RF magnetron sputtering. *Surf. Interface Anal.* **2002**, *34* (1), 732–734.

51. Ding, X. Z.; Zeng, X. T.; Xie, H., Cubic boron nitride films deposited by unbalanced RF magnetron sputtering and pulsed DC substrate bias. *Thin Solid Films.* **2003**, *429* (1), 22–27.

52. Caicedo, J. M.; Bejarano, G.; Zambrano, G.; Baca, E.; Morán, O.; Prieto, P., Cubic and hexagonal boron-nitride (c-BN/h-BN) thin films deposited *in situ* by RF magnetron sputtering. *Phys. Status Solidi B.* **2005**, *242* (9), 1920–1923.

53. Wiggins, M. D.; Aita, C. R.; Hickernell, F. S., Radio frequency sputter deposited boron nitride films. *J. Vac. Sci. Technol. A.* **1984**, *2* (2), 322–325.

54. Bewilogua, K.; Buth, J.; Hübsch, H.; Grischke, M., Preparation of c-BN containing films by reactive RF sputtering. *Diamond Relat. Mater.* **1993**, *2* (8), 1206–1210.

55. Lüthje, H.; Bewilogua, K.; Daaud, S.; Johansson, M.; Hultman, L., Preparation of cubic boron nitride films by use of electrically conductive boron carbide targets. *Thin Solid Films.* **1995**, *257* (1), 40–45.

56. Johansson, M. P.; Hultman, L.; Daaud, S.; Bewilogua, K.; Lüthje, H.; Schütze, A.; Kouptsidis, S.; Theunissen, G. S. A. M., Microstructure of BN:C films deposited on Si substrates by reactive sputtering from a B4C target. *Thin Solid Films.* **1996**, *287* (1), 193–201.

57. Yang, T.-S.; Tsai, T.-H.; Lee, C.-H.; Cheng, C.-L.; Wong, M.-S., Deposition of carbon-containing cubic boron nitride films by pulsed-DC magnetron sputtering. *Thin Solid Films.* **2001**, *398–399*, 285–290.

58. Park, J.-K.; Ko, J.-S.; Baik, Y.-J., Effect of hydrogen addition on the residual stress of B–C–N films with cubic boron nitride phase prepared by RF magnetron sputtering of a B4C target. *Surf. Coat. Technol.* **2013**, *215*, 104–109.

59. Kim, H. S.; Park, J. K.; Lee, W. S.; Baik, Y. J., Variation of residual stress in cubic boron nitride film caused by hydrogen addition during unbalanced magnetron sputtering. *Thin Solid Films.* **2011**, *519* (22), 7871–7874.

60. Hahn, J.; Friedrich, M.; Pintaske, R.; Schaller, M.; Kahl, N.; Zahn, D. R. T.; Richter, F., Cubic boron nitride films by DC and RF magnetron sputtering: layer characterization and process diagnostics. *Diamond Relat. Mater.* **1996**, *5* (10), 1103–1112.

61. Matsumoto, S.; Sato, Y.; Kamo, M.; Setaka, N., Vapor deposition of diamond particles from methane. *Jpn. J. Appl. Phys.* **1982**, *21* (Part 2, No. 4), L183–L185.

62. Matsumoto, S.; Sato, Y.; Tsutsumi, M.; Setaka, N., Growth of diamond particles from methane-hydrogen gas. *J. Mater. Sci.* **1982**, *17* (11), 3106–3112.

63. Kamo, M.; Sato, Y.; Matsumoto, S.; Setaka, N., Diamond synthesis from gas phase in microwave plasma. *J. Cryst. Growth.* **1983**, *62* (3), 642–644.

64. Angus, J. C.; Hayman, C. C., Low-pressure, metastable growth of diamond and diamondlike and phases. *Science.* **1988**, *241* (4868), 913.

65. Szmidt, J.; Jakubowski, A.; Michalski, A.; Rusek, A., Basic properties of metal/insulator/semiconductor structures containing borazone and diamond layers produced by the reactive pulse plasma method. *Thin Solid Films.* **1983**, *110* (1), 7–20.

66. Chayahara, A.; Yokoyama, H.; Imura, T.; Osaka, Y.; Fujisawa, M., Function of substrate bias potential for formation of cubic boron nitride films in plasma CVD technique. *Jpn. J. Appl. Phys.* **1987**, *26* (9A), L1435.

67. Chayahara, A.; Yokoyama, H.; Imura, T.; Osaka, Y.; Fujisawa, M., Properties of BN thin films deposited by plasma CVD. *Appl. Surf. Sci.* **1988**, *33*, 561–566.

68. Chayahara, A.; Yokoyama, H.; Imura, T.; Osaka, Y.; Fujisawa, M., Reflectance spectra of BN materials in the vacuum ultraviolet. *Jpn. J. Appl. Phys.* **1988**, *27* (3R), 440.

69. Yokoyama, H.; Okamoto, M.; Hamada, T.; Imura, T.; Osaki, Y.; Chayahara, A.; Fujisawa, M., Imaginary part of the dielectric function of sintered and microcrystalline cubic boron nitride. *Jpn. J. Appl. Phys.* **1989**, *28* (3R), 555.

70. Okamoto, M.; Yokoyama, H.; Osaka, Y., Formation of cubic boron nitride film on Si with boron buffer layers. *Jpn. J. Appl. Phys.* **1990**, *29* (5R), 930.

71. Osaka, Y.; Chayahara, A.; Yokoyama, H.; Okamoto, M.; Hamada, T.; Imura, T.; Fujisawa, M., Characterization of BN thin films deposited by plasma CVD. *Mater. Sci. Forum.* **1991**, *54–55*, 277–294.

72. Dworschak, W.; Jung, K.; Ehrhardt, H., Growth of cubic boron nitride coatings in a magnetic field enhanced RF glow discharge. *Diamond Relat. Mater.* **1994**, *3* (4), 337–340.

73. Dworschak, W.; Jung, K.; Ehrhardt, H., Growth mechanism of cubic boron nitride in a RF glow discharge. *Thin Solid Films.* **1995**, *254* (1), 65–74.

74. Ichiki, T.; Momose, T.; Yoshida, T., Effects of the substrate bias on the formation of cubic boron nitride by inductively coupled plasma enhanced chemical vapor deposition. *J. Appl. Phys.* **1994**, *75* (3), 1330–1334.

75. Ichiki, T.; Yoshida, T., Preparation of cubic boron nitride films by low pressure inductively coupled plasma enhanced chemical vapor deposition. *Appl. Phys. Lett.* **1994**, *64* (7), 851–853.

76. Berns, D. H.; Cappelli, M. A., Cubic boron nitride synthesis in low-density supersonic plasma flows. *Appl. Phys. Lett.* **1996**, *68* (19), 2711–2713.

77. Berns, D. H.; Cappelli, M. A., Low energy ion impact-enhanced growth of cubic boron nitride in a supersonic nitrogen/argon plasma flow. *J. Mater. Res.* **1997**, *12* (8), 2014–2026.

78. Karim, M. Z.; Cameron, D. C.; Murphy, M. J.; Hashmi, M. S. J., Plasma deposition of cubic boron nitride films from non-toxic material at low temperatures. *Surf. Coat. Technol.* **1991**, *49* (1), 416–421.

79. Karim, M. Z.; Cameron, D. C.; Hashmi, M. S. J., Vapour deposited boron nitride thin films. *Mater. Des.* **1992**, *13* (4), 207–214.

80. Karim, M. Z.; Cameron, D. C.; Hashmi, M. S. J., Plasma diagnostics in the growth of c-BN films. *Diamond Relat. Mater.* **1994**, *3* (4), 551–554.

81. Amagi, S.; Takahashi, D.; Yoshida, T., Threshold sheath potential for the nucleation and growth of cubic boron nitride by inductively coupled plasma enhanced chemical-vapor deposition. *Appl. Phys. Lett.* **1997**, *70* (8), 946–948.

82. Saitoh, H.; Hirose, T.; Hirotsu, Y.; Ichinose, Y., Synthesis of BN films by the plasma CVD with various solids: BH_3NH_3, H_3BO_3, and $NaBH_4$. *Surf. Coat. Technol.* **1989**, *39*, 265–273.

83. Saitoh, H.; Yarbrough, W. A., Preparation and characterization of nanocrystalline cubic boron nitride by microwave plasma-enhanced chemical vapor deposition. *Appl. Phys. Lett.* **1991**, *58* (20), 2228–2230.

84. Saitoh, H.; Yarbrough, W. A., Growth of cubic boron nitride on diamond particles by microwave plasma enhanced chemical vapor deposition. *Appl. Phys. Lett.* **1991**, *58* (22), 2482–2484.

85. Ichinose, Y.; Saitoh, H.; Hirotsu, Y., Synthesis of cubic BN from the gas phase by a new plasma chemical vapour deposition method using RF waves and a tungsten filament. *Surf. Coat. Technol.* **1990**, *43*, 116–127.

86. Loeffler, J.; Konyashin, I.; Bill, J.; Uhlig, H.; Aldinger, F., PACVD nano-crystalline BCN thin films obtained by use of an organoboron precursor. *Diamond Relat. Mater.* **1997**, *6* (5), 608–611.

87. Loh, K. P.; Sakaguchi, I.; Nishitani-Gamo, M.; Taniguchi, T.; Ando, T., Hydrogen-induced surface structuring of a cubic boron nitride (100) face studied by low-energy electron diffraction and electron spectroscopic techniques. *Phys. Rev. B.* **1998**, *57* (12), 7266–7274.

88. Zhang, F.; Guo, Y.; Song, Z.; Chen, G., Deposition of high quality cubic boron nitride films on nickel substrates. *Appl. Phys. Lett.* **1994**, *65* (8), 971–973.

89. Song, Z.; Zhang, F.; Guo, Y.; Chen, G., Textured growth of cubic boron nitride film on nickel substrates. *Appl. Phys. Lett.* **1994**, *65* (21), 2669–2671.

90. Guo, Y.; Song, Z.; Zhang, Y.; Zhang, F.; Chen, G., Growth of cubic boron nitride films by hot filament assisted RF plasma CVD. *Phys. Status Solidi A.* **1994**, *143* (1), K13–K17.

91. Shapoval, S. Y.; Petrashov, V. T.; Popov, O. A.; Westner, A. O.; Yoder, M. D.; Lok, C. K. C., Cubic boron nitride films deposited by electron cyclotron resonance plasma. *Appl. Phys. Lett.* **1990**, *57* (18), 1885–1886.

92. Kalss, W.; Haubner, R.; Lux, B., Preparation of BN films in the B–N–F system. *Diamond Relat. Mater.* **1998**, *7* (2), 369–375.

93. Zhang, W. J.; Matsumoto, S., The roles of hydrogen and fluorine in the deposition of cubic boron nitride films in the Ar–N_2–BF_3–H_2 system. *Chem. Phys. Lett.* **2000**, *330* (3–4), 243–248.

94. Zhang, W. J.; Matsumoto, S., The effects of DC bias voltage on the crystal size and crystal quality of cBN films. *Appl. Phys. A.* **2000**, *71* (4), 469–472.

95. Matsumoto, S.; Zhang, W., High-rate deposition of high-quality, thick cubic boron nitride films by bias-assisted DC jet plasma chemical vapor deposition. *Jpn. J. Appl. Phys.* **2000**, *39* (5B), L442.

96. Matsumoto, S.; Zhang, W. J., Synthesis of boron nitride films by microwave plasma chemical vapor deposition in fluorine-containing gases. *Jpn. J. Appl. Phys.* **2001**, *40* (6A), L570.

97. Zhang, W. J.; Jiang, X.; Matsumoto, S., High-quality, faceted cubic boron nitride films grown by chemical vapor deposition. *Appl. Phys. Lett.* **2001**, *79* (27), 4530–4532.

98. Matsumoto, S.; Zhang, W. J., The introducing of fluorine into the deposition of BN: A successful method to obtain high-quality, thick cBN films with low residual stress. *Diamond Relat. Mater.* **2001**, *10* (9–10), 1868–1874.

99. Zhang, W. J.; Chan, C. Y.; Chan, K. M.; Bello, I.; Lifshitz, Y.; Lee, S. T., Deposition of large-area, high-quality cubic boron nitride films by ECR-enhanced microwave-plasma CVD. *Appl. Phys. A.* **2003**, *76* (6), 953–955.

100. Chan, C. Y.; Zhang, W. J.; Matsumoto, S.; Bello, I.; Lee, S. T., A nanoindentation study of thick cBN films prepared by chemical vapor deposition. *J. Cryst. Growth.* **2003**, *247* (3–4), 438–444.

101. Chan, C. Y.; Zhang, W. J.; Meng, X. M.; Chan, K. M.; Bello, I.; Lifshitz, Y.; Lee, S. T., The growth of thick cBN films employing fluorine chemistry and ECR deposition. *Diamond Relat. Mater.* **2003**, *12* (3–7), 1162–1168.

102. Zhang, W. J.; Matsumoto, S.; Li, Q.; Bello, I.; Lee, S. T., Growth behavior of cubic boron nitride films in a two-step process: Changing bias voltage, gas composition, and substrate temperature. *Adv. Funct. Mater.* **2002**, *12* (4), 250–254.

103. Yu, J.; Zheng, Z.; Ong, H. C.; Wong, K. Y.; Matsumoto, S.; Lau, W. M., Thermal stability of cubic boron nitride films deposited by chemical vapor deposition. *J. Phys. Chem. B.* **2006**, *110* (42), 21073–21076.

104. Chong, Y. M.; Leung, K. M.; Ma, K. L.; Zhang, W. J.; Bello, I.; Lee, S. T., Growing cubic boron nitride films at different temperatures. *Diamond Relat. Mater.* **2006**, *15* (4–8), 1155–1160.

105. Zhang, W.; Bello, I.; Lifshitz, Y.; Chan, K. M.; Meng, X.; Wu, Y.; Chan, C. Y.; Lee, S. T., Epitaxy on diamond by chemical vapor deposition: A route to high-quality cubic boron nitride for electronic applications. *Adv. Mater.* **2004**, *16* (16), 1405–1408.

106. Zhang, W. J.; Bello, I.; Lifshitz, Y.; Chan, K. M.; Wu, Y.; Chan, C. Y.; Meng, X. M.; Lee, S. T., Thick and adherent cubic boron nitride films grown on diamond interlayers by fluorine-assisted chemical vapor deposition. *Appl. Phys. Lett.* **2004**, *85* (8), 1344–1346.

107. Zhang, X. W.; Boyen, H. G.; Deyneka, N.; Ziemann, P.; Banhart, F.; Schreck, M., Epitaxy of cubic boron nitride on (001)-oriented diamond. *Nat. Mater.* **2003**, *2* (5), 312–315.

108. Zhang, W. J.; Chan, C. Y.; Meng, X. M.; Fung, M. K.; Bello, I.; Lifshitz, Y.; Lee, S. T.; Jiang, X., The mechanism of chemical vapor deposition of cubic boron nitride films from fluorine-containing species. *Angew. Chem. Int. Ed.* **2005**, *44* (30), 4749–4753.

109. Larsson, K.; Lunell, S.; Carlsson, J. O., Adsorption of hydrocarbons on a diamond (111) surface: An *ab initio* quantum-mechanical study. *Phys. Rev. B.* **1993**, *48* (4), 2666–2674.

110. Larsson, K.; Carlsson, J. O.; Lunell, S., Nearest-neighbor influence on hydrocarbon adsorption on diamond (111) studied by *ab initio* calculations. *Phys. Rev. B.* **1995**, *51* (15), 10003–10012.

111. Larsson, K.; Carlsson, J. O., Surface processes in cubic boron nitride growth: A theoretical study. *J. Phys. Chem. B.* **1999**, *103* (31), 6533–6538.

112. Larsson, K.; Carlsson, J. O., Surface processes in cubic boron nitride growth: A theoretical study. *J. Phys. Chem. B.* **1999**, *103* (31), 6533–6538.

113. Larsson, K., CVD growth of cubic boron nitride: A theoretical/experimental approach. *Thin Solid Films.* **2006**, *515* (2), 401–406.

114. Mårlid, B.; Larsson, K.; Carlsson, J. O., Theoretical investigation of hydrogen- and halogen-terminated c-BN (111) clusters. *Phys. Rev. B.* **1999**, *60* (23), 16065–16072.

115. Mårlid, B.; Larsson, K.; Carlsson, J. O. Proceedings of the second symposium on III-V nitride materials and processes, Pennington, NJ, Abernathy, C. A. (Ed.), The Electrochemical Society: Pennington, NJ, 1999, 151–158.

116. Zhang, R. Q.; Zhang, D.; Zhao, Y. L.; Lee, S. T., Fluorination induced etching selectivity of boron nitride phases. *J. Phys. Chem. B.* **2004**, *108* (23), 7597–7602.

117. Olander, J.; Larsson, K., Initial growth of hexagonal and cubic boron nitride: A theoretical study. *Phys. Rev. B.* **2003**, *68* (7), 075411.

118. Olander, J.; Larsson, K., Cubic boron nitride growth from NH_3 and BBr_3 precursors: A theoretical study. *Diamond Relat. Mater.* **2002**, *11* (3), 1286–1289.

119. Larsson, K.; Carlsson, J.-O., Surface migration during diamond growth studied by molecular orbital calculations. *Phys. Rev. B.* **1999**, *59* (12), 8315–8322.

120. Karlsson, J.; Larsson, K., Adsorption of growth species on the c-BN (100) surface. *J. Phys. Chem. C.* **2011**, *115* (34), 16977–16983.

121. Basche, M., Boron nitride coating and a process of producing the same. Google Patents. 1964.

122. Rand, M. J.; Roberts, J. F., Preparation and properties of thin film boron nitride. *J. Electrochem. Soc.* **1968**, *115* (4), 423–429.

123. Gebhardt, J. J., CVD boron nitride infiltration of fibrous structures: Properties of low temperature. Electrochemical Society Inc.: United States, 1973.

124. Pierson, H. O., Boron nitride composites by chemical vapor deposition. *J. Compos. Mater.* **1975**, *9* (3), 228–240.

125. Clerc, G.; Gerlach, P. In *Pyrolytic Boron Nitride*, 5th Int'l Conf on Chemical Vapour Deposition, UK, 1975, 777–785.

126. Sano, M.; Aoki, M., Chemical vapour deposition of thin films of BN onto fused silica and sapphire. *Thin Solid Films.* **1981**, *83* (2), 247–251.

127. Motojima, S.; Tamura, Y.; Sugiyama, K., Low temperature deposition of hexagonal BN films by chemical vapour deposition. *Thin Solid Films.* **1982**, *88* (3), 269–274.

128. Hannache, H.; Naslain, R.; Bernard, C., Boron nitride chemical vapour infiltration of fibrous materials from BCl_3, NH_3, H_2 or BF_3, NH_3 mixtures: A thermodynamic and experimental approach. *J. Less. Comm. Met.* **1983**, *95* (2), 221–246.

129. Matsuda, T.; Uno, N.; Nakae, H.; Hirai, T., Synthesis and structure of chemically vapour-deposited boron nitride. *J. Mater. Sci.* **1986**, *21* (2), 649–658.

130. Hurd, J. L.; Perry, D. L.; Lee, B. T.; Yu, K. M.; Bourret, E. D.; Haller, E. E., Polycrystalline hexagonal boron nitride films on SiO_2 for III–V semiconductor applications. *J. Mater. Res.* **1989**, *4* (02), 350–354.

131. Tanji, H.; Suzuki, M.; Nomura, K.; Watanabe, S.; Kawasaki, T., Pyrolytic boron nitride article and method for producing the same. Google Patents. 1986.

132. Arya, S. P. S.; D'Amico, A., Preparation, properties and applications of boron nitride thin films. *Thin Solid Films.* **1988**, *157* (2), 267–282.

133. Takahashi, T.; Itoh, H.; Takeuchi, A., Chemical vapor deposition of hexagonal boron nitride thick film on iron. *J. Cryst. Growth.* **1979**, *47* (2), 245–250.

134. Takahashi, T.; Itoh, H.; Kuroda, M., Structure and properties of CVD-BN thick film prepared on carbon steel substrate. *J. Cryst. Growth.* **1981**, *53* (2), 418–422.

135. Takahashi, T.; Itoh, H.; Ohtake, A., Chemical vapor deposition of boron nitride films on nickel substrate. *J. Ceram. Soc. Japan.* **1981**, *89* (1026), 63–68.

136. Andújar, J. L.; Bertran, E.; Maniette, Y., Microstructure of highly oriented, hexagonal, boron nitride thin films grown on crystalline silicon by radio frequency plasma-assisted chemical vapor deposition. *J. Appl. Phys.* **1996**, *80* (11), 6553–6555.

137. Sugino, T.; Kawasaki, S.; Tanioka, K.; Shirafuji, J., Electron emission from boron nitride coated Si field emitters. *Appl. Phys. Lett.* **1997**, *71* (18), 2704–2706.

138. Nagashima, A.; Tejima, N.; Gamou, Y.; Kawai, T.; Oshima, C., Electronic dispersion relations of monolayer hexagonal boron nitride formed on the Ni(111) surface. *Phys. Rev. B.* **1995**, *51* (7), 4606–4613.

139. Nagashima, A.; Tejima, N.; Gamou, Y.; Kawai, T.; Oshima, C., Electronic structure of monolayer hexagonal boron nitride physisorbed on metal surfaces. *Phys. Rev. Lett.* **1995**, *75* (21), 3918–3921.

140. Nagashima, A.; Tejima, N.; Gamou, Y.; Kawai, T.; Oshima, C., Electronic states of monolayer hexagonal boron nitride formed on the metal surfaces. *Surf. Sci.* **1996**, *357*, 307–311.

141. Chuhei, O.; Ayato, N., Ultra-thin epitaxial films of graphite and hexagonal boron nitride on solid surfaces. *J. Phys.: Condens. Matter.* **1997**, *9* (1), 1.

142. Desrosiers, R. M.; Greve, D. W.; Gellman, A. J., Nucleation of boron nitride thin films on Ni(100). *Surf. Sci.* **1997**, *382* (1), 35–48.

143. Auwärter, W.; Kreutz, T. J.; Greber, T.; Osterwalder, J., XPD and STM investigation of hexagonal boron nitride on Ni(111). *Surf. Sci.* **1999**, *429* (1–3), 229–236.

144. Gamou, Y.; Terai, M.; Nagashima, A.; Oshima, C., Atomic Structural analysis of a monolayer epitaxial film of hexagonal boron nitride/ Ni(111) studied by LEED intensity analysis (interfaces by various techniques). *Sci. Rep.* **1997**, *RITU A44*, 211.

145. Auwärter, W.; Muntwiler, M.; Osterwalder, J.; Greber, T., Defect lines and two-domain structure of hexagonal boron nitride films on Ni(111). *Surf. Sci.* **2003**, *545* (1–2), L735–L740.

146. Muntwiler, M.; Auwärter, W.; Baumberger, F.; Hoesch, M.; Thomas, G.; Osterwalder, J., Determining adsorbate structures from substrate emission X-ray photoelectron diffraction. *Surf. Sci.* **2001**, *472* (1–2), 125–132.

147. Osterwalder, J.; Auwärter, W. M., M.; Greber, T., Growth morphologies and defect structure in hexagonal boron nitride films on Ni(111): A combined STM and XPD study. *e-J. Surf. Sci.* **2003**, *1*, 124–129.

148. Greber, T.; Brandenberger, L.; Corso, M.; Tamai, A.; Osterwalder, J.; uuml; rg, Single layer hexagonal boron nitride films on Ni(110). *e-J. Surf. Sci.* **2006**, *4*, 410–413.

149. Shi, Y.; Hamsen, C.; Jia, X.; Kim, K. K.; Reina, A.; Hofmann, M.; Hsu, A. L.; Zhang, K.; Li, H.; Juang, Z.-Y.; Dresselhaus, M. S.; Li, L.-J.; Kong, J., Synthesis of few-layer hexagonal boron nitride thin film by chemical vapor deposition. *Nano Lett.* **2010**, *10* (10), 4134–4139.

150. Song, L.; Ci, L.; Lu, H.; Sorokin, P. B.; Jin, C.; Ni, J.; Kvashnin, A. G.; Kvashnin, D. G.; Lou, J.; Yakobson, B. I.; Ajayan, P. M., Large scale growth and characterization of atomic hexagonal boron nitride layers. *Nano Lett.* **2010**, *10* (8), 3209–3215.

151. Preobrajenski, A. B.; Vinogradov, A. S.; Mårtensson, N., Monolayer of h-BN chemisorbed on Cu(111) and Ni(111): The role of the transition metal 3d states. *Surf. Sci.* **2005**, *582* (1–3), 21–30.

152. Chatterjee, S.; Luo, Z.; Acerce, M.; Yates, D. M.; Johnson, A. T. C.; Sneddon, L. G., Chemical vapor deposition of boron nitride nanosheets on metallic substrates *via* decaborane/ammonia reactions. *Chem. Mater.* **2011**, *23* (20), 4414–4416.

153. Kim, K. K.; Hsu, A.; Jia, X.; Kim, S. M.; Shi, Y.; Dresselhaus, M.; Palacios, T.; Kong, J., Synthesis and characterization of hexagonal

boron nitride film as a dielectric layer for graphene devices. *ACS Nano.* **2012**, *6* (10), 8583–8590.

154. Kim, K. K.; Hsu, A.; Jia, X.; Kim, S. M.; Shi, Y.; Hofmann, M.; Nezich, D.; Rodriguez-Nieva, J. F.; Dresselhaus, M.; Palacios, T.; Kong, J., Synthesis of monolayer hexagonal boron nitride on Cu foil using chemical vapor deposition. *Nano Lett.* **2012**, *12* (1), 161–166.
155. Lee, K. H.; Shin, H.-J.; Lee, J.; Lee, I.-y.; Kim, G.-H.; Choi, J.-Y.; Kim, S.-W., Large-scale synthesis of high-quality hexagonal boron nitride nanosheets for large-area graphene electronics. *Nano Lett.* **2012**, *12* (2), 714–718.
156. Ning, G.; Jinquan, W.; Lili, F.; Yi, J.; Dayao, L.; Hongwei, Z.; Kunlin, W.; Dehai, W., Controllable growth of triangular hexagonal boron nitride domains on copper foils by an improved low-pressure chemical vapor deposition method. *Nanotechnology.* **2012**, *23* (41), 415605.
157. Tay, R. Y.; Wang, X.; Tsang, S. H.; Loh, G. C.; Singh, R. S.; Li, H.; Mallick, G.; Tong Teo, E. H., A systematic study of the atmospheric pressure growth of large-area hexagonal crystalline boron nitride film. *J. Mater. Chem. C.* **2014**, *2* (9), 1650–1657.
158. Wang, M.; Kim, M.; Odkhuu, D.; Park, N.; Lee, J.; Jang, W.-J.; Kahng, S.-J.; Ruoff, R. S.; Song, Y. J.; Lee, S., Catalytic transparency of hexagonal boron nitride on copper for chemical vapor deposition growth of large-area and high-quality graphene. *ACS Nano.* **2014**, *8* (6), 5478–5483.
159. Stehle, Y.; Meyer, H. M.; Unocic, R. R.; Kidder, M.; Polizos, G.; Datskos, P. G.; Jackson, R.; Smirnov, S. N.; Vlassiouk, I. V., Synthesis of hexagonal boron nitride monolayer: Control of nucleation and crystal morphology. *Chem. Mater.* **2015**, *27* (23), 8041–8047.
160. Kidambi, P. R.; Blume, R.; Kling, J.; Wagner, J. B.; Baehtz, C.; Weatherup, R. S.; Schloegl, R.; Bayer, B. C.; Hofmann, S., *in situ* observations during chemical vapor deposition of hexagonal boron nitride on polycrystalline copper. *Chem. Mater.* **2014**, *26* (22), 6380–6392.
161. Tay, R. Y.; Li, H.; Tsang, S. H.; Zhu, M.; Loeblein, M.; Jing, L.; Leong, F. N.; Teo, E. H. T., Trimethylamine borane: A new single-source precursor for monolayer h-BN single crystals and h-BCN thin films. *Chem. Mater.* **2016**, *28* (7), 2180–2190.
162. Koepke, J. C.; Wood, J. D.; Chen, Y.; Schmucker, S. W.; Liu, X.; Chang, N. N.; Nienhaus, L.; Do, J. W.; Carrion, E. A.; Hewaparakrama, J.; Rangarajan, A.; Datye, I.; Mehta, R.; Haasch, R. T.; Gruebele, M.; Girolami, G. S.; Pop, E.; Lyding, J. W., Role of pressure in the growth

of hexagonal boron nitride thin films from ammonia-borane. *Chem. Mater.* **2016,** *28* (12), 4169–4179.

163. Song, X.; Li, Q.; Ji, J.; Gu, Y.; Huo, C.; Zou, Y.; Zhi, C.; Zeng, H., A comprehensive investigation on CVD growth thermokinetics of h-BN white graphene. *2D Mater.* **2016,** *3* (3), 035007.

164. McLean, B.; Webber, G. B.; Page, A. J., Boron nitride nanotube nucleation *via* network fusion during catalytic chemical vapor deposition. *J. Am. Chem. Soc.* **2019,** *141* (34), 13385–13393.

165. Lu, G.; Wu, T.; Yuan, Q.; Wang, H.; Wang, H.; Ding, F.; Xie, X.; Jiang, M., Synthesis of large single-crystal hexagonal boron nitride grains on Cu–Ni alloy. *Nat. Commun.* **2015,** *6,* 6160.

166. Ismach, A.; Chou, H.; Ferrer, D. A.; Wu, Y.; McDonnell, S.; Floresca, H. C.; Covacevich, A.; Pope, C.; Piner, R.; Kim, M. J.; Wallace, R. M.; Colombo, L.; Ruoff, R. S., Toward the controlled synthesis of hexagonal boron nitride films. *ACS Nano.* **2012,** *6* (7), 6378–6385.

167. Lu, G.; Wu, T.; Yang, P.; Yang, Y.; Jin, Z.; Chen, W.; Jia, S.; Wang, H.; Zhang, G.; Sun, J.; Ajayan, P. M.; Lou, J.; Xie, X.; Jiang, M., Synthesis of high-quality graphene and hexagonal boron nitride monolayer in-plane heterostructure on Cu–Ni Alloy. *Adv. Sci.* **2017,** *4* (9), 1700076.

168. Uchida, Y.; Nakandakari, S.; Kawahara, K.; Yamasaki, S.; Mitsuhara, M.; Ago, H., Controlled growth of large-area uniform multilayer hexagonal boron nitride as an effective 2D substrate. *ACS Nano.* **2018,** *12* (6), 6236–6244.

169. Azizi, A.; AlSaud, M. A.; Alem, N., Controlled growth and atomic-scale characterization of two-dimensional hexagonal boron nitride crystals. *J. Cryst. Growth.* **2018,** *496–497,* 51–56.

170. Chang, R.-J.; Wang, X.; Wang, S.; Sheng, Y.; Porter, B.; Bhaskaran, H.; Warner, J. H., Growth of large single-crystalline monolayer hexagonal boron nitride by oxide-assisted chemical vapor deposition. *Chem. Mater.* **2017,** *29* (15), 6252–6260.

171. Ji, Y.; Calderon, B.; Han, Y.; Cueva, P.; Jungwirth, N. R.; Alsalman, H. A.; Hwang, J.; Fuchs, G. D.; Muller, D. A.; Spencer, M. G., Chemical vapor deposition growth of large single-crystal mono-, bi-, tri-layer hexagonal boron nitride and their interlayer stacking. *ACS Nano.* **2017,** *11* (12), 12057–12066.

172. Li, M.; Zhou, S.; Wang, R.; Yu, Y.; Wong, H.; Luo, Z.; Li, H.; Gan, L.; Zhai, T., *in situ* formed nanoparticle-assisted growth of large-size single crystalline h-BN on copper. *Nanoscale.* **2018,** *10* (37), 17865–17872.

173. Tay, R. Y.; Park, H. J.; Ryu, G. H.; Tan, D.; Tsang, S. H.; Li, H.; Liu, W.; Teo, E. H. T.; Lee, Z.; Lifshitz, Y.; Ruoff, R. S., Synthesis of aligned symmetrical multifaceted monolayer hexagonal boron nitride single crystals on resolidified copper. *Nanoscale.* **2016,** *8* (4), 2434–2444.

174. Wang, L.; Wu, B.; Liu, H.; Huang, L.; Li, Y.; Guo, W.; Chen, X.; Peng, P.; Fu, L.; Yang, Y.; Hu, P.; Liu, Y., Water-assisted growth of large-sized single crystal hexagonal boron nitride grains. *Mater. Chem. Front.* **2017,** *1* (9), 1836–1840.

175. Yin, J.; Yu, J.; Li, X.; Li, J.; Zhou, J.; Zhang, Z.; Guo, W., Large single-crystal hexagonal boron nitride monolayer domains with controlled morphology and straight merging boundaries. *Small.* **2015,** *11* (35), 4497–4502.

176. Song, X.; Gao, J.; Nie, Y.; Gao, T.; Sun, J.; Ma, D.; Li, Q.; Chen, Y.; Jin, C.; Bachmatiuk, A.; Rümmeli, M. H.; Ding, F.; Zhang, Y.; Liu, Z., Chemical vapor deposition growth of large-scale hexagonal boron nitride with controllable orientation. *Nano Res.* **2015,** *8* (10), 3164–3176.

177. Wang, L.; Xu, X.; Zhang, L.; Qiao, R.; Wu, M.; Wang, Z.; Zhang, S.; Liang, J.; Zhang, Z.; Zhang, Z.; Chen, W.; Xie, X.; Zong, J.; Shan, Y.; Guo, Y.; Willinger, M.; Wu, H.; Li, Q.; Wang, W.; Gao, P.; Wu, S.; Zhang, Y.; Jiang, Y.; Yu, D.; Wang, E.; Bai, X.; Wang, Z.-J.; Ding, F.; Liu, K., Epitaxial growth of a 100-square-centimetre single-crystal hexagonal boron nitride monolayer on copper. *Nature.* **2019,** *570* (7759), 91–95.

178. Meng, J.; Ming, B.; Zhang, X.; Gao, M.; Cheng, L.; Yin, Z.; Wang, D.; Li, X.; You, J.; Wang, R., Controlled growth of unidirectionally aligned hexagonal boron nitride domains on single crystal Ni (111)/MgO thin films. *Cryst. Growth Des.* **2019,** *19* (1), 453–459.

179. Corso, M.; Auwärter, W.; Muntwiler, M.; Tamai, A.; Greber, T.; Osterwalder, J., Boron nitride nanomesh. *Science.* **2004,** *303* (5655), 217.

180. Berner, S.; Corso, M.; Widmer, R.; Groening, O.; Laskowski, R.; Blaha, P.; Schwarz, K.; Goriachko, A.; Over, H.; Gsell, S.; Schreck, M.; Sachdev, H.; Greber, T.; Osterwalder, J., Boron nitride nanomesh: Functionality from a corrugated monolayer. *Angew. Chem.* **2007,** *119* (27), 5207–5211.

181. Goriachko, A.; He; Knapp, M.; Over, H.; Corso, M.; Brugger, T.; Berner, S.; Osterwalder, J.; Greber, T., Self-assembly of a hexagonal boron nitride nanomesh on Ru(0001). *Langmuir.* **2007,** *23* (6), 2928–2931.

182. Morscher, M.; Corso, M.; Greber, T.; Osterwalder, J., Formation of single layer h-BN on Pd(111). *Surf. Sci.* **2006,** *600* (16), 3280–3284.

183. Corso, M.; Greber, T.; Osterwalder, J., h-BN on Pd(110): A tunable system for self-assembled nanostructures. *Surf. Sci.* **2005**, *577* (2–3), L78–L84.

184. Müller, F.; Hüfner, S.; Sachdev, H., Epitaxial growth of boron nitride on a Rh(111) multilayer system: Formation and fine tuning of a BN-nanomesh. *Surf. Sci.* **2009**, *603* (3), 425–432.

185. Sachdev, H.; Müller, F.; Hüfner, S., BN analogues of graphene: On the formation mechanism of boronitrene layers—solids with extreme structural anisotropy. *Diamond Relat. Mater.* **2010**, *19* (7–9), 1027–1033.

186. Preobrajenski, A. B.; Vinogradov, A. S.; Ng, M. L.; Ćavar, E.; Westerström, R.; Mikkelsen, A.; Lundgren, E.; Mårtensson, N., Influence of chemical interaction at the lattice-mismatched h-BN/Rh(111) and h-BN/Pt(111) interfaces on the overlayer morphology. *Phys. Rev. B.* **2007**, *75* (24), 245412.

187. Müller, F.; Hüfner, S.; Sachdev, H.; Laskowski, R.; Blaha, P.; Schwarz, K., Epitaxial growth of hexagonal boron nitride on Ag(111). *Phys. Rev. B.* **2010**, *82* (11), 113406.

188. Müller, F.; Stöwe, K.; Sachdev, H., Symmetry versus commensurability: Epitaxial growth of hexagonal boron nitride on Pt(111) from B-trichloroborazine (ClBNH)3. *Chem. Mater.* **2005**, *17* (13), 3464–3467.

189. Kim, G.; Jang, A. R.; Jeong, H. Y.; Lee, Z.; Kang, D. J.; Shin, H. S., Growth of high-crystalline, single-layer hexagonal boron nitride on recyclable platinum foil. *Nano Lett.* **2013**, *13* (4), 1834–1839.

190. Gao, Y.; Ren, W.; Ma, T.; Liu, Z.; Zhang, Y.; Liu, W.-B.; Ma, L.-P.; Ma, X.; Cheng, H.-M., Repeated and controlled growth of monolayer, bilayer and few-layer hexagonal boron nitride on Pt foils. *ACS Nano.* **2013**, *7* (6), 5199–5206.

191. Park, J.-H.; Park, J. C.; Yun, S. J.; Kim, H.; Luong, D. H.; Kim, S. M.; Choi, S. H.; Yang, W.; Kong, J.; Kim, K. K.; Lee, Y. H., Large-area monolayer hexagonal boron nitride on Pt foil. *ACS Nano.* **2014**, *8* (8), 8520–8528.

192. Park, J.-H.; Choi, S. H.; Zhao, J.; Song, S.; Yang, W.; Kim, S. M.; Kim, K. K.; Lee, Y. H., Thickness-controlled multilayer hexagonal boron nitride film prepared by plasma-enhanced chemical vapor deposition. *Curr. Appl. Phys.* **2016**, *16* (9), 1229–1235.

193. Grad, G. B.; Blaha, P.; Schwarz, K.; Auwärter, W.; Greber, T., Density functional theory investigation of the geometric and spintronic structure of h-BN/Ni(111) in view of photoemission and STM experiments. *Phys. Rev. B.* **2003**, *68* (8), 085404.

194. Che, J. G.; Cheng, H.-P., First-principles investigation of a monolayer of C60 on h-BN/Ni(111). *Phys. Rev. B.* **2005**, *72* (11), 115436.

195. Huda, M. N.; Kleinman, L., h-BN monolayer adsorption on the Ni(111) surface: A density functional study. *Phys. Rev. B.* **2006**, *74*, 075418.

196. Sun, X.; Pratt, A.; Li, Z. Y.; Ohtomo, M.; Sakai, S.; Yamauchi, Y., The adsorption of h-BN monolayer on the Ni(111) surface studied by density functional theory calculations with a semiempirical long-range dispersion correction. *J. Appl. Phys.* **2014**, *115* (17), 17C117.

197. Liu, L.; Siegel, D. A.; Chen, W.; Liu, P.; Guo, J.; Duscher, G.; Zhao, C.; Wang, H.; Wang, W.; Bai, X.; McCarty, K. F.; Zhang, Z.; Gu, G., Unusual role of epilayer–substrate interactions in determining orientational relations in van der Waals epitaxy. *PNAS.* **2014**, *111* (47), 16670–16675.

198. Laskowski, R.; Blaha, P.; Gallauner, T.; Schwarz, K., Single-layer model of the hexagonal boron nitride nanomesh on the Rh(111) surface. *Phys. Rev. Lett.* **2007**, *98* (10), 106802.

199. Laskowski, R.; Blaha, P., *ab initio* study of h-BN nanomeshes on Ru(001), Rh(111), and Pt(111). *Phys. Rev. B.* **2010**, *81* (7), 075418.

200. Laskowski, R.; Blaha, P.; Schwarz, K., Bonding of hexagonal BN to transition metal surfaces: An *ab initio* density-functional theory study. *Phys. Rev. B.* **2008**, *78* (4), 045409.

201. Liu, Y.; Bhowmick, S.; Yakobson, B. I., BN white graphene with "colorful" edges: The energies and morphology. *Nano Lett.* **2011**, *11* (8), 3113–3116.

202. Zhao, R.; Gao, J.; Liu, Z.; Ding, F., The reconstructed edges of the hexagonal BN. *Nanoscale.* **2015**, *7* (21), 9723–9730.

203. Fu, X.; Zhang, R., Energetics of hexagonal boron nitride nanostructures: Edge dependence and truncation effects. *Nanoscale.* **2017**, *9* (20), 6734–6740.

204. Zhao, R.; Li, F.; Liu, Z.; Liu, Z.; Ding, F., The transition metal surface passivated edges of hexagonal boron nitride (h-BN) and the mechanism of h-BN's chemical vapor deposition (CVD) growth. *Phys. Chem. Chem. Phys.* **2015**, *17* (43), 29327–29334.

205. Zhu, H.; Zhao, X.; Li, H.; Zhao, R., Revealing stable geometries and magic clusters of hexagonal boron nitride in the nucleation of chemical vapor deposition growth on Ni(111)/Cu(111) surfaces: A theoretical study. *Phys. Chem. Chem. Phys.* **2020**, *22* (7), 4023–4031.

206. Zhang, Z.; Liu, Y.; Yang, Y.; Yakobson, B. I., Growth mechanism and morphology of hexagonal boron nitride. *Nano Lett.* **2016**, *16* (2), 1398–1403.

207. Liu, S.; van Duin, A. C. T.; van Duin, D. M.; Liu, B.; Edgar, J. H., Atomistic insights into nucleation and formation of hexagonal boron nitride on nickel from first-principles-based reactive molecular dynamics simulations. *ACS Nano.* **2017**, *11* (4), 3585–3596.

208. Liu, S.; Comer, J.; van Duin, A. C. T.; van Duin, D. M.; Liu, B.; Edgar, J. H., Predicting the preferred morphology of hexagonal boron nitride domain structure on nickel from ReaxFF-based molecular dynamics simulations. *Nanoscale.* **2019**, *11* (12), 5607–5616.

209. Liu, Z.; Dong, J.; Ding, F., Geometry of hexagonal boron nitride clusters in the initial stages of chemical vapor deposition growth on Cu(111) surface. *Nanoscale.* **2019**.

210. Novoselov, K. S.; Geim, A. K.; Morozov, S. V.; Jiang, D.; Zhang, Y.; Dubonos, S. V.; Grigorieva, I. V.; Firsov, A. A., Electric field effect in atomically thin carbon films. *Science.* **2004**, *306* (5696), 666.

211. Pacilé, D.; Meyer, J. C.; Girit, Ç. Ö.; Zettl, A., The two-dimensional phase of boron nitride: Few-atomic-layer sheets and suspended membranes. *Appl. Phys. Lett.* **2008**, *92* (13), 133107.

212. Alem, N.; Erni, R.; Kisielowski, C.; Rossell, M. D.; Gannett, W.; Zettl, A., Atomically thin hexagonal boron nitride probed by ultrahigh-resolution transmission electron microscopy. *Phys. Rev. B.* **2009**, *80* (15), 155425.

213. Watanabe, K.; Taniguchi, T.; Kanda, H., Direct-bandgap properties and evidence for ultraviolet lasing of hexagonal boron nitride single crystal. *Nat. Mater.* **2004**, *3* (6), 404–409.

214. An, L.; Yu, Y.; Bai, C.; Bai, Y.; Zhang, B.; Gao, K.; Wang, X.; Lai, Z.; Zhang, J., Simultaneous production and functionalization of hexagonal boron nitride nanosheets by solvent-free mechanical exfoliation for superlubricant water-based lubricant additives. *NPJ 2D Mater. Appl.* **2019**, *3* (1), 28.

215. Han, W.-Q.; Wu, L.; Zhu, Y.; Watanabe, K.; Taniguchi, T., Structure of chemically derived mono- and few-atomic-layer boron nitride sheets. *Appl. Phys. Lett.* **2008**, *93* (22), 223103.

216. Zhi, C.; Bando, Y.; Tang, C.; Kuwahara, H.; Golberg, D., Large-scale fabrication of boron nitride nanosheets and their utilization in polymeric composites with improved thermal and mechanical properties. *Adv. Mater.* **2009**, *21* (28), 2889–2893.

217. Zhao, L.; Yan, L.; Wei, C.; Wang, Z.; Jia, L.; Ran, Q.; Huang, X.; Ren, J., Aqueous-phase exfoliation and functionalization of boron nitride nanosheets using tannic acid for thermal management applications. *Ind. Eng. Chem. Res.* **2020**, *59* (37), 16273–16282.

218. Sun, W.; Meng, Y.; Fu, Q.; Wang, F.; Wang, G.; Gao, W.; Huang, X.; Lu, F., High-yield production of boron nitride nanosheets and its uses as a catalyst support for hydrogenation of nitroaromatics. *ACS Appl. Mater. Int.* **2016**, *8* (15), 9881–9888.

219. Zhang, B.; Wu, Q.; Yu, H.; Bulin, C.; Sun, H.; Li, R.; Ge, X.; Xing, R., High-efficient liquid exfoliation of boron nitride nanosheets using aqueous solution of alkanolamine. *Nanoscale Res. Lett.* **2017**, *12* (1), 596.

220. Ye, H.; Lu, T.; Xu, C.; Han, B.; Meng, N.; Xu, L., Liquid-phase exfoliation of hexagonal boron nitride into boron nitride nanosheets in common organic solvents with hyperbranched polyethylene as stabilizer. *Macromolecular Chemistry and Physics.* **2018**, *219* (6), 1700482.

221. Kulkarni, H. B.; Tambe, P. B.; Joshi, G. M., Influence of surfactant assisted exfoliation of hexagonal boron nitride nanosheets on mechanical, thermal and dielectric properties of epoxy nanocomposites. *Compos. Interfaces.* **2020**, *27* (6), 529–550.

222. Cheng, Z.-L.; Ma, Z.-S.; Ding, H.-L.; Liu, Z., Environmentally friendly, scalable exfoliation for few-layered hexagonal boron nitride nanosheets (BNNSs) by multi-time thermal expansion based on released gases. *J. Mater. Chem. C.* **2019**, *7* (46), 14701–14708.

223. Deshmukh, A. R.; Jeong, J. W.; Lee, S. J.; Park, G. U.; Kim, B. S., Ultrasound-assisted facile green synthesis of hexagonal boron nitride nanosheets and their applications. *ACS Sust. Chem. Eng.* **2019**, *7* (20), 17114–17125.

224. Coleman, J. N.; Lotya, M.; O'Neill, A.; Bergin, S. D.; King, P. J.; Khan, U.; Young, K.; Gaucher, A.; De, S.; Smith, R. J.; Shvets, I. V.; Arora, S. K.; Stanton, G.; Kim, H.-Y.; Lee, K.; Kim, G. T.; Duesberg, G. S.; Hallam, T.; Boland, J. J.; Wang, J. J.; Donegan, J. F.; Grunlan, J. C.; Moriarty, G.; Shmeliov, A.; Nicholls, R. J.; Perkins, J. M.; Grieveson, E. M.; Theuwissen, K.; McComb, D. W.; Nellist, P. D.; Nicolosi, V., Two-dimensional nanosheets produced by liquid exfoliation of layered materials. *Science.* **2011**, *331* (6017), 568.

225. Zhou, K.-G.; Mao, N.-N.; Wang, H.-X.; Peng, Y.; Zhang, H.-L., A Mixed-solvent strategy for efficient exfoliation of inorganic graphene analogues. *Angew. Chem. Int. Ed.* **2011**, *50* (46), 10839–10842.

226. Cao, L.; Emami, S.; Lafdi, K., Large-scale exfoliation of hexagonal boron nitride nanosheets in liquid phase. *Mater. Exp.* **2014**, *4* (2), 165–171.

227. Nie, X.; Li, G.; Jiang, Z.; Li, W.; Ouyang, T.; Wang, J., Co-solvent exfoliation of hexagonal boron nitride: Effect of raw bulk boron nitride size and co-solvent composition. *Nanomaterials.* **2020**, *10* (6).

228. Shen, J.; He, Y.; Wu, J.; Gao, C.; Keyshar, K.; Zhang, X.; Yang, Y.; Ye, M.; Vajtai, R.; Lou, J.; Ajayan, P. M., Liquid phase exfoliation of two-dimensional materials by directly probing and matching surface tension components. *Nano Lett.* **2015**, *15* (8), 5449–5454.

229. Wang, H.; Su, X.; Song, T.; Li, Z.; Zhao, Y.; Lou, H.; Wang, J., Scalable exfoliation and dispersion of few-layer hexagonal boron nitride nanosheets in NMP-salt solutions. *Appl. Surf. Sci.* **2019**, *488*, 656–661.

230. Wang, S.; Tao, B.; Yu, S.; Wei, C.; Zhou, T.; Chen, X.; Han, C.; Wang, C., Insight into the liquid-phase exfoliation to prepare BN nanosheets. *Mater. Lett.* **2020**, *269*, 127644.

231. Lin, Y.; Williams, T. V.; Xu, T.-B.; Cao, W.; Elsayed-Ali, H. E.; Connell, J. W., Aqueous dispersions of few-layered and monolayered hexagonal boron nitride nanosheets from sonication-assisted hydrolysis: Critical role of water. *J. Phys. Chem. C.* **2011**, *115* (6), 2679–2685.

232. Mukhopadhyay, T. K.; Datta, A., Deciphering the role of solvents in the liquid phase exfoliation of hexagonal boron nitride: A molecular dynamics simulation study. *J. Phys. Chem. C.* **2017**, *121* (1), 811–822.

233. Wang, X.; Zhi, C.; Li, L.; Zeng, H.; Li, C.; Mitome, M.; Golberg, D.; Bando, Y., "Chemical blowing" of thin-walled bubbles: High-throughput fabrication of large-area, few-layered BN and Cx-BN nanosheets. *Adv. Mater.* **2011**, *23* (35), 4072–4076.

234. Lee, D.; Lee, B.; Park, K. H.; Ryu, H. J.; Jeon, S.; Hong, S. H., Scalable exfoliation process for highly soluble boron nitride nanoplatelets by hydroxide-assisted ball milling. *Nano Lett.* **2015**, *15* (2), 1238–1244.

235. Fu, L.; Lai, G.; Chen, G.; Lin, C.-T.; Yu, A., Microwave irradiation-assisted exfoliation of boron nitride nanosheets: A platform for loading high density of nanoparticles. *ChemistrySelect.* **2016**, *1* (8), 1799–1803.

236. Liu, Z.; Wang, Y.; Wang, Z.; Yao, Y.; Dai, J.; Das, S.; Hu, L., Solvo-thermal microwave-powered two-dimensional material exfoliation. *Chem. Commun.* **2016**, *52* (33), 5757–5760.

237. Andriani, Y.; Song, J.; Lim, P. C.; Seng, D. H. L.; Lai, D. M. Y.; Teo, S. L.; Kong, J.; Wang, X.; Zhang, X.; Liu, S., Green and efficient production of boron nitride nanosheets *via* oxygen doping-facilitated liquid exfoliation. *Ceram. Int.* **2019**, *45* (4), 4909–4917.

238. Rafiei-Sarmazdeh, Z.; Jafari, S. H.; Ahmadi, S. J.; Zahedi-Dizaji, S. M., Large-scale exfoliation of hexagonal boron nitride with combined fast quenching and liquid exfoliation strategies. *J. Mater. Sci.* **2016**, *51* (6), 3162–3169.

239. Ma, Z.-S.; Ding, H.-L.; Liu, Z.; Cheng, Z.-L., Preparation and tribological properties of hydrothermally exfoliated ultrathin hexagonal boron nitride nanosheets (BNNSs) in mixed NaOH/KOH solution. *J. Alloys Compd.* **2019**, *784*, 807–815.

240. Wang, N.; Yang, G.; Wang, H.; Yan, C.; Sun, R.; Wong, C.-P., A universal method for large-yield and high-concentration exfoliation of two-dimensional hexagonal boron nitride nanosheets. *Mater. Today.* **2019**, *27*, 33–42.

241. Sokołowska, A.; Rudnicki, J.; Kostecki, M.; Wojtkiewicz, S.; Sawosz, P.; Chodun, R.; Zdunek, K.; Olszyna, A., Electric field used as the substitute for ultrasounds in the liquid exfoliation of hexagonal boron nitride. *Microelectron. Eng.* **2014**, *126*, 124–128.

242. Zhao, G.; Wu, Y.; Shao, Y.; Hao, X., Large-quantity and continuous preparation of two-dimensional nanosheets. *Nanoscale.* **2016**, *8* (10), 5407–5411.

243. Gonzalez Ortiz, D.; Pochat-Bohatier, C.; Cambedouzou, J.; Bechelany, M.; Miele, P., Exfoliation of hexagonal boron nitride (h-BN) in liquid phase by ion intercalation. *Nanomaterials.* **2018**, *8* (9).

244. Gao, W.; Zhao, Y.; Yin, H., Lateral size selection of liquid exfoliated hexagonal boron nitride nanosheets. *RSC Adv.* **2018**, *8* (11), 5976–5983.

245. Gissler, W.; Haupt, J.; Crabb, T. A.; Gibson, P. N.; Rickerby, D. G., Evidence for mixed-phase nanocrystalline boron nitride films. *Mater. Sci. Eng. A.* **1991**, *139*, 284–289.

246. Jäger, S.; Bewilogua, K.; Klages, C. P., Infrared spectroscopic investigations on h-BN and mixed h/c-BN thin films. *Thin Solid Films.* **1994**, *245* (1), 50–54.

247. Sutter, P.; Lahiri, J.; Zahl, P.; Wang, B.; Sutter, E., Scalable synthesis of uniform few-layer hexagonal boron nitride dielectric films. *Nano Lett.* **2013**, *13* (1), 276–281.

248. Peng, C.; Yin, S.; Zhang, D.; Chao, X.; Quarterman, P.; Wang, J.-P., Investigation on the structural property of the sputtered hcp-phase boron nitride tunnel barrier for spintronic applications. *AIP Adv.* **2020**, *10* (3), 035005.

249. Stewart, D. M.; Lad, R. J., Enhanced crystallinity of h-BN films induced by substrate bias during magnetron sputtering. *Phys. Status Solidi B.* **2018**, *255* (3), 1700458.

250. Ohta, J.; Fujioka, H., Sputter synthesis of wafer-scale hexagonal boron nitride films *via* interface segregation. *APL Mater.* **2017**, *5* (7), 076107.

251. Meng, J.; Zhang, X.; Wang, Y.; Yin, Z.; Liu, H.; Xia, J.; Wang, H.; You, J.; Jin, P.; Wang, D.; Meng, X.-M., Aligned growth of millimeter-size hexagonal boron nitride single-crystal domains on epitaxial nickel thin film. *Small.* **2017**, *13* (18), 1604179.

252. Paisley, M. J.; Sitar, Z.; Yan, B.; Davis, R. F., Growth of boron nitride films by gas molecular-beam epitaxy. *J. Vac. Sci. Technol. B.* **1990**, *8* (2), 323–326.

253. Tsai, C. L.; Kobayashi, Y.; Akasaka, T.; Kasu, M., Molecular beam epitaxial growth of hexagonal boron nitride on Ni(111) substrate. *J. Cryst. Growth.* **2009**, *311* (10), 3054–3057.

254. Nakhaie, S.; Wofford, J. M.; Schumann, T.; Jahn, U.; Ramsteiner, M.; Hanke, M.; Lopes, J. M. J.; Riechert, H., Synthesis of atomically thin hexagonal boron nitride films on nickel foils by molecular beam epitaxy. *Appl. Phys. Lett.* **2015**, *106* (21), 213108.

255. Courtade, E.; Han, B.; Nakhaie, S.; Robert, C.; Marie, X.; Renucci, P.; Taniguchi, T.; Watanabe, K.; Geelhaar, L.; Lopes, J. M. J.; Urbaszek, B., Spectrally narrow exciton luminescence from monolayer MoS_2 and $MoSe_2$ exfoliated onto epitaxially grown hexagonal BN. *Appl. Phys. Lett.* **2018**, *113* (3), 032106.

256. Wofford, J. M.; Nakhaie, S.; Krause, T.; Liu, X.; Ramsteiner, M.; Hanke, M.; Riechert, H.; J. Lopes, J. M., A hybrid MBE-based growth method for large-area synthesis of stacked hexagonal boron nitride/graphene heterostructures. *Sci. Rep.* **2017**, *7* (1), 43644.

257. Tian, H.; He, Y.; Das, P.; Cui, Z.; Shi, W.; Khanaki, A.; Lake, R. K.; Liu, J., Growth dynamics of millimeter-sized single-crystal hexagonal boron

nitride monolayers on secondary recrystallized Ni (100) substrates. *Advanced Materials Interfaces.* **2019**, *6* (22), 1901198.

258. Nakhaie, S.; Heilmann, M.; Krause, T.; Hanke, M.; Lopes, J. M. J., Nucleation and growth of atomically thin hexagonal boron nitride on Ni/MgO(111) by molecular beam epitaxy. *J. Appl. Phys.* **2019**, *125* (11), 115301.

259. Tonkikh, A. A.; Voloshina, E. N.; Werner, P.; Blumtritt, H.; Senkovskiy, B.; Güntherodt, G.; Parkin, S. S. P.; Dedkov, Y. S., Structural and electronic properties of epitaxial multilayer h-BN on Ni(111) for spintronics applications. *Sci. Rep.* **2016**, *6* (1), 23547.

260. Driver, M. S.; Beatty, J. D.; Olanipekun, O.; Reid, K.; Rath, A.; Voyles, P. M.; Kelber, J. A., Atomic layer epitaxy of h-BN(0001) multilayers on Co(0001) and molecular beam epitaxy growth of graphene on h-BN(0001)/Co(0001). *Langmuir.* **2016**, *32* (11), 2601–2607.

261. Xu, Z.; Tian, H.; Khanaki, A.; Zheng, R.; Suja, M.; Liu, J., Large-area growth of multi-layer hexagonal boron nitride on polished cobalt foils by plasma-assisted molecular beam epitaxy. *Sci. Rep.* **2017**, *7* (1), 43100.

262. Cheng, T. S.; Summerfield, A.; Mellor, C. J.; Khlobystov, A. N.; Eaves, L.; Foxon, C. T.; Beton, P. H.; Novikov, S. V., High-temperature molecular beam epitaxy of hexagonal boron nitride with high active nitrogen fluxes. *Materials.* **2018**, *11*.

263. Xu, Z.; Khanaki, A.; Tian, H.; Zheng, R.; Suja, M.; Zheng, J.-G.; Liu, J., Direct growth of hexagonal boron nitride/graphene heterostructures on cobalt foil substrates by plasma-assisted molecular beam epitaxy. *Appl. Phys. Lett.* **2016**, *109* (4), 043110.

264. Cho, Y.-J.; Summerfield, A.; Davies, A.; Cheng, T. S.; Smith, E. F.; Mellor, C. J.; Khlobystov, A. N.; Foxon, C. T.; Eaves, L.; Beton, P. H.; Novikov, S. V., Hexagonal boron nitride tunnel barriers grown on graphite by high temperature molecular beam epitaxy. *Sci. Rep.* **2016**, *6* (1), 34474.

265. Cheng, T. S.; Summerfield, A.; Mellor, C. J.; Davies, A.; Khlobystov, A. N.; Eaves, L.; Foxon, C. T.; Beton, P. H.; Novikov, S. V., High-temperature molecular beam epitaxy of hexagonal boron nitride layers. *J. Vac. Sci. Technol. B.* **2018**, *36* (2), 02D103.

266. Yang, X.; Pristovsek, M.; Nitta, S.; Liu, Y.; Honda, Y.; Koide, Y.; Kawarada, H.; Amano, H., Epitaxial combination of two-dimensional hexagonal boron nitride with single-crystalline diamond substrate. *ACS Appl. Mater. Int.* **2020**, *12* (41), 46466–46475.

267. Yang, X.; Nitta, S.; Pristovsek, M.; Liu, Y.; Liao, Y.; Kushimoto, M.; Honda, Y.; Amano, H., Scalable synthesis of multilayer h-BN on AlN by metalorganic vapor phase epitaxy: Nucleation and growth mechanism. *2D Mater.* **2019**, *7* (1), 015004.

268. Laleyan, D. A.; Mengle, K.; Zhao, S.; Wang, Y.; Kioupakis, E.; Mi, Z., Effect of growth temperature on the structural and optical properties of few-layer hexagonal boron nitride by molecular beam epitaxy. *Opt. Express.* **2018**, *26* (18), 23031–23039.

269. Kobayashi, Y.; Akasaka, T., Hexagonal BN epitaxial growth on (0 0 0 1) sapphire substrate by MOVPE. *J. Cryst. Growth.* **2008**, *310* (23), 5044–5047.

270. Kobayashi, Y.; Akasaka, T.; Makimoto, T., Hexagonal boron nitride grown by MOVPE. *J. Cryst. Growth.* **2008**, *310* (23), 5048–5052.

271. Kobayashi, Y.; Nakamura, T.; Akasaka, T.; Makimoto, T.; Matsumoto, N., Hexagonal boron nitride on Ni(111) substrate grown by flow-rate modulation epitaxy. *J. Cryst. Growth.* **2007**, *298*, 325–327.

272. Heilmann, M.; Bashouti, M.; Riechert, H.; Lopes, J. M. J., Defect mediated van der Waals epitaxy of hexagonal boron nitride on graphene. *2D Mater.* **2018**, *5* (2), 025004.

273. Hamilton, E. J. M.; Dolan, S. E.; Mann, C. M.; Colijn, H. O.; McDonald, C. A.; Shore, S. G., Preparation of amorphous boron nitride and its conversion to a turbostratic, tubular form. *Science.* **1993**, *260* (5108), 659.

274. Shelimov, K. B.; Moskovits, M., Composite nanostructures based on template-grown boron nitride nanotubules. *Chem. Mater.* **2000**, *12* (1), 250–254.

275. Hamilton, E. J. M.; Dolan, S. E.; Mann, C. M.; Colijn, H. O.; Shore, S. G., Preparation of amorphous boron nitride from the reaction of haloborazines with alkali metals and formation of a novel tubular morphology by thermal annealing. *Chem. Mater.* **1995**, *7* (1), 111–117.

276. Hamilton, E. J. M.; Boyd, E. P.; Shore, S. G., Comments on the transformation of amorphous boron nitride to a tubular morphology. *Chem. Mater.* **2001**, *13* (4), 1428–1428.

277. Rubio, A.; Corkill, J. L.; Cohen, M. L., Theory of graphitic boron nitride nanotubes. *Phys. Rev. B.* **1994**, *49* (7), 5081–5084.

278. Iijima, S., Helical microtubules of graphitic carbon. *Nature.* **1991**, *354* (6348), 56–58.

279. Krätschmer, W.; Lamb, L. D.; Fostiropoulos, K.; Huffman, D. R., Solid C_{60}: A new form of carbon. *Nature.* **1990,** *347* (6291), 354–358.

280. Chopra, N. G.; Luyken, R.; Cherrey, K.; Crespi, V. H., Boron nitride nanotubes. *Science.* **1995,** *269* (5226), 966–967.

281. Narita, I.; Oku, T., Synthesis of boron nitride nanotubes by using YB6 powder. *Solid State Commun.* **2002,** *122* (9), 465–468.

282. Loiseau, A.; Willaime, F.; Demoncy, N.; Hug, G.; Pascard, H., Boron nitride nanotubes with reduced numbers of layers synthesized by arc discharge. *Phys. Rev. Lett.* **1996,** *76* (25), 4737–4740.

283. Yahachi, S.; Masatomo, M.; Takehisa, M., Structures of boron nitride nanotubes with single-layer and multilayers produced by arc discharge. *Jpn. J. Appl. Phys.* **1999,** *38* (1R), 159.

284. Terrones, M.; Hsu, W. K.; Terrones, H.; Zhang, J. P.; Ramos, S.; Hare, J. P.; Castillo, R.; Prassides, K.; Cheetham, A. K.; Kroto, H. W.; Walton, D. R. M., Metal particle catalysed production of nanoscale BN structures. *Chem. Phys. Lett.* **1996,** *259* (5), 568–573.

285. Saito, Y.; Maida, M., Square, pentagon, and heptagon rings at BN nanotube tips. *J. Phys. Chem. A.* **1999,** *103* (10), 1291–1293.

286. Cumings, J.; Zettl, A., Mass-production of boron nitride double-wall nanotubes and nanococoons. *Chem. Phys. Lett.* **2000,** *316* (3–4), 211–216.

287. Altoe, M. V. P.; Sprunck, J. P.; Gabriel, J. C. P.; Bradley, K., Nanococoon seeds for BN nanotube growth. *J. Mater. Sci.* **2003,** *38* (24), 4805–4810.

288. Yeh, Y.-W.; Raitses, Y.; Koel, B. E.; Yao, N., Stable synthesis of few-layered boron nitride nanotubes by anodic arc discharge. *Sci. Rep.* **2017,** *7* (1), 3075.

289. Fathalizadeh, A.; Pham, T.; Mickelson, W.; Zettl, A., Scaled synthesis of boron nitride nanotubes, nanoribbons, and nanococoons using direct feedstock injection into an extended-pressure, inductively-coupled thermal plasma. *Nano Lett.* **2014,** *14* (8), 4881–4886.

290. Kim, K. S.; Kingston, C. T.; Hrdina, A.; Jakubinek, M. B.; Guan, J.; Plunkett, M.; Simard, B., Hydrogen-catalyzed, pilot-scale production of small-diameter boron nitride nanotubes and their macroscopic assemblies. *ACS Nano.* **2014,** *8* (6), 6211–6220.

291. Santra, B.; Ko, H.-Y.; Yeh, Y.-W.; Martelli, F.; Kaganovich, I.; Raitses, Y.; Car, R., Root-growth of boron nitride nanotubes: Experiments and *ab initio* simulations. *Nanoscale.* **2018,** *10* (47), 22223–22230.

292. Kim, M.; Lee, Y. H.; Oh, J.-H.; Hong, S.-H.; Min, B.-I.; Kim, T.-H.; Choi, S., Synthesis of boron nitride nanotubes using triple DC thermal plasma reactor with hydrogen injection. *Chem. Eng. J.* **2020**, *395*, 125148.

293. Golberg, D.; Bando, Y.; Eremets, M.; Takemura, K.; Kurashima, K.; Yusa, H., Nanotubes in boron nitride laser heated at high pressure. *Appl. Phys. Lett.* **1996**, *69* (14), 2045–2047.

294. Thess, A.; Lee, R.; Nikolaev, P.; Dai, H.; Petit, P.; Robert, J.; Xu, C.; Lee, Y. H.; Kim, S. G.; Rinzler, A. G.; Colbert, D. T.; Scuseria, G. E.; Tománek, D.; Fischer, J. E.; Smalley, R. E., Crystalline ropes of metallic carbon nanotubes. *Science.* **1996**, *273* (5274), 483.

295. Yu, D. P.; Sun, X. S.; Lee, C. S.; Bello, I.; Lee, S. T.; Gu, H. D.; Leung, K. M.; Zhou, G. W.; Dong, Z. F.; Zhang, Z., Synthesis of boron nitride nanotubes by means of excimer laser ablation at high temperature. *Appl. Phys. Lett.* **1998**, *72* (16), 1966–1968.

296. Kim, J. H.; Cho, H.; Pham, T. V.; Hwang, J. H.; Ahn, S.; Jang, S. G.; Lee, H.; Park, C.; Kim, C. S.; Kim, M. J., Dual growth mode of boron nitride nanotubes in high temperature pressure laser ablation. *Sci. Rep.* **2019**, *9* (1), 15674.

297. Arenal, R.; Stephan, O.; Cochon, J.-L.; Loiseau, A., Root-growth mechanism for single-walled boron nitride nanotubes in laser vaporization technique. *J. Am. Chem. Soc.* **2007**, *129* (51), 16183–16189.

298. Lee, R. S.; Gavillet, J.; Chapelle, M. L. d. l.; Loiseau, A.; Cochon, J. L.; Pigache, D.; Thibault, J.; Willaime, F., Catalyst-free synthesis of boron nitride single-walled nanotubes with a preferred zig-zag configuration. *Phys. Rev. B.* **2001**, *64* (12), 121405.

299. Laude, T.; Matsui, Y.; Marraud, A.; Jouffrey, B., Long ropes of boron nitride nanotubes grown by a continuous laser heating. *Appl. Phys. Lett.* **2000**, *76* (22), 3239–3241.

300. Zhou, G. W.; Zhang, Z.; Bai, Z. G.; Yu, D. P., Catalyst effects on formation of boron nitride nano-tubules synthesized by laser ablation. *Solid State Commun.* **1999**, *109* (8), 555–559.

301. Golberg, D.; Rode, A.; Bando, Y.; Mitome, M.; Gamaly, E.; Luther-Davies, B., Boron nitride nanostructures formed by ultra-high-repetition rate laser ablation. *Diamond Relat. Mater.* **2003**, *12* (8), 1269–1274.

302. Naumov, V. G.; Kosyrev, F. K.; Vostrikov, V. G.; Arutyunyan, N. R.; Obraztsova, E. D.; Konov, V. I.; Jiang, H.; Nasibulin, A.; Kauppinen, E., Synthesis of boron nitride multi-walled nanotubes by laser ablation technique. *Laser Physics.* **2009**, *19* (5), 1198–1200.

303. Han, W.; Bando, Y.; Kurashima, K.; Sato, T., Synthesis of boron nitride nanotubes from carbon nanotubes by a substitution reaction. *Appl. Phys. Lett.* **1998**, *73* (21), 3085–3087.

304. Han, W.-Q.; Mickelson, W.; Cumings, J.; Zettl, A., Transformation of BxCyNz nanotubes to pure BN nanotubes. *Appl. Phys. Lett.* **2002**, *81* (6), 1110–1112.

305. Golberg, D.; Bando, Y.; Kurashima, K.; Sato, T., MoO_3-promoted synthesis of multi-walled BN nanotubes from C nanotube templates. *Chem. Phys. Lett.* **2000**, *323* (1–2), 185–191.

306. Bechelany, M.; Brioude, A.; Stadelmann, P.; Bernard, S.; Cornu, D.; Miele, P., Preparation of BN microtubes/nanotubes with a unique chemical process. *J. Phys. Chem. C.* **2008**, *112* (47), 18325–18330.

307. Chen, Y.; Chadderton, L. T.; Gerald, J. F.; Williams, J. S., A solid-state process for formation of boron nitride nanotubes. *Appl. Phys. Lett.* **1999**, *74* (20), 2960–2962.

308. Chen, Y.; Fitz Gerald, J.; Williams, J. S.; Bulcock, S., Synthesis of boron nitride nanotubes at low temperatures using reactive ball milling. *Chem. Phys. Lett.* **1999**, *299* (3–4), 260–264.

309. Chen, Y.; Conway, M.; Williams, J. S.; Zou, J., Large-quantity production of high-yield boron nitride nanotubes. *J. Mater. Res.* **2002**, *17* (08), 1896–1899.

310. Chen, Y.; Zou, J.; Campbell, S. J.; Le Caer, G., Boron nitride nanotubes: Pronounced resistance to oxidation. *Appl. Phys. Lett.* **2004**, *84* (13), 2430–2432.

311. Li, Y.; Zhou, J. E.; Zhao, K.; Tung, S.; Schneider, E., Synthesis of boron nitride nanotubes from boron oxide by ball milling and annealing process. *Mater. Lett.* **2009**, *63* (20), 1733–1736.

312. Kim, J.; Lee, S.; Uhm, Y. R.; Jun, J.; Rhee, C. K.; Kim, G. M., Synthesis and growth of boron nitride nanotubes by a ball milling–annealing process. *Acta Mater.* **2011**, *59* (7), 2807–2813.

313. Lim, S. H.; Luo, J.; Ji, W.; Lin, J., Synthesis of boron nitride nanotubes and its hydrogen uptake. *Catal. Today.* **2007**, *120* (3), 346–350.

314. Wang, S.; Wen, D.; Wang, G.; Yang, L.; Guo, P.; Qiao, Y., Multitudinous industrialized fabrication of BNNT with high-energy ball milling and arc-discharge-aided template methods. *J. Nanopart. Res.* **2017**, *19* (10), 352.

315. Lourie, O. R.; Jones, C. R.; Bartlett, B. M.; Gibbons, P. C.; Ruoff, R. S.; Buhro, W. E., CVD growth of boron nitride nanotubes. *Chem. Mater.* **2000**, *12* (7), 1808–1810.

316. Loiseau, A.; Willaime, F.; Demoncy, N.; Schramchenko, N.; Hug, G.; Colliex, C.; Pascard, H., Boron nitride nanotubes. *Carbon.* **1998**, *36* (5), 743–752.

317. Tang, C.; Bando, Y.; Sato, T., Catalytic growth of boron nitride nanotubes. *Chem. Phys. Lett.* **2002**, *362* (3–4), 185–189.

318. Fu, J. J.; Lu, Y. N.; Xu, H.; Huo, K. F.; Wang, X. Z.; Li, L.; Hu, Z.; Chen, Y., The synthesis of boron nitride nanotubes by an extended vapour–liquid–solid method. *Nanotechnology.* **2004**, *15* (7), 727–730.

319. Golberg, D.; Bando, Y.; Kurashima, K.; Sato, T., Ropes of BN multi-walled nanotubes. *Solid State Commun.* **2000**, *116* (1), 1–6.

320. Golberg, D.; Bando, Y.; Bourgeois, L.; Kurashima, K.; Sato, T., Insights into the structure of BN nanotubes. *Appl. Phys. Lett.* **2000**, *77* (13), 1979–1981.

321. Golberg, D.; Bando, Y., Unique morphologies of boron nitride nanotubes. *Appl. Phys. Lett.* **2001**, *79* (3), 415–417.

322. Golberg, D.; Bando, Y.; Kurashima, K.; Sato, T., Synthesis and characterization of ropes made of BN multiwalled nanotubes. *Scr. Mater.* **2001**, *44* (8), 1561–1565.

323. Tang, C. C.; Lamy de la Chapelle, M.; Li, P.; Liu, Y. M.; Dang, H. Y.; Fan, S. S., Catalytic growth of nanotube and nanobamboo structures of boron nitride. *Chem. Phys. Lett.* **2001**, *342* (5–6), 492–496.

324. Saner Okan, B.; Kocabaş, Z. Ö.; Nalbant Ergün, A.; Baysal, M.; Letofsky-Papst, I.; Yürüm, Y., Effect of reaction temperature and catalyst type on the formation of boron nitride nanotubes by chemical vapor deposition and measurement of their hydrogen storage capacity. *Ind. Eng. Chem. Res.* **2012**, *51* (35), 11341–11347.

325. Li, J.; Li, J.; Yin, Y.; Chen, Y.; Bi, X., Water-assisted chemical vapor deposition synthesis of boron nitride nanotubes and their photoluminescence property. *Nanotechnology.* **2013**, *24* (36), 365605.

326. Wang, X. Z.; Wu, Q.; Hu, Z.; Chen, Y., Template-directed synthesis of boron nitride nanotube arrays by microwave plasma chemical reaction. *Electrochim. Acta.* **2007**, *52* (8), 2841–2844.

327. Guo, L.; Singh, R. N., Selective growth of boron nitride nanotubes by plasma-enhanced chemical vapor deposition at low substrate temperature. *Nanotechnology.* **2008**, *19* (6), 065601.

328. Guo, L.; Singh, R. N., Catalytic growth of boron nitride nanotubes using gas precursors. *Physica E.* **2009**, *41* (3), 448–453.

329. Su, C.-Y.; Chu, W.-Y.; Juang, Z.-Y.; Chen, K.-F.; Cheng, B.-M.; Chen, F.-R.; Leou, K.-C.; Tsai, C.-H., Large-scale synthesis of boron nitride nanotubes with iron-supported catalysts. *J. Phys. Chem. C.* **2009**, *113* (33), 14732–14738.

330. Tang, C.; Bando, Y.; Sato, T.; Kurashima, K., A novel precursor for synthesis of pure boron nitride nanotubes. *Chem. Commun.* **2002**, *12* (12), 1290–1291.

331. Tang, C.; Bando, Y.; Golberg, D., Multi-walled BN nanotubes synthesized by carbon-free method. *J. Solid State Chem.* **2004**, *177* (8), 2670–2674.

332. Li, J.; Lin, H.; Chen, Y.; Su, Q.; Huang, Q., The effect of iron oxide on the formation of boron nitride nanotubes. *Chem. Eng. J.* **2011**, *174* (2), 687–692.

333. Özmen, D.; Sezgi, N. A.; Balcı, S., Synthesis of boron nitride nanotubes from ammonia and a powder mixture of boron and iron oxide. *Chem. Eng. J.* **2013**, *219*, 28–36.

334. Nithya, J. S. M.; Pandurangan, A., Efficient mixed metal oxide routed synthesis of boron nitride nanotubes. *RSC Adv.* **2014**, *4* (51), 26697–26705.

335. Wang, J.; Kayastha, V. K.; Yap, Y. K.; Fan, Z.; Lu, J. G.; Pan, Z.; Ivanov, I. N.; Puretzky, A. A.; Geohegan, D. B., Low temperature growth of boron nitride nanotubes on substrates. *Nano Lett.* **2005**, *5* (12), 2528–2532.

336. Baysal, M.; Bilge, K.; Yıldızhan, M. M.; Yorulmaz, Y.; Öncel, Ç.; Papila, M.; Yürüm, Y., Catalytic synthesis of boron nitride nanotubes at low temperatures. *Nanoscale.* **2018**, *10* (10), 4658–4662.

337. Matveev, A. T.; Firestein, K. L.; Steinman, A. E.; Kovalskii, A. M.; Lebedev, O. I.; Shtansky, D. V.; Golberg, D., Boron nitride nanotube growth *via* boron oxide assisted chemical vapor transport-deposition process using LiNO3 as a promoter. *Nano Res.* **2015**, *8* (6), 2063–2072.

338. E, S.; Wu, L.; Li, C.; Zhu, Z.; Long, X.; Geng, R.; Zhang, J.; Li, Z.; Lu, W.; Yao, Y., Growth of boron nitride nanotubes from magnesium diboride catalysts. *Nanoscale.* **2018**, *10* (29), 13895–13901.

339. Wang, L.; Han, D.; Luo, J.; Li, T.; Lin, Z.; Yao, Y., Highly efficient growth of boron nitride nanotubes and the thermal conductivity of their polymer composites. *J. Phys. Chem. C.* **2018**, *122* (3), 1867–1873.

340. Abbas, S.; Abbas, A.; Huang, Y.; Lin, J.; Liu, Z.; Fang, Y.; Tang, C., Synthesis of boron nitride nanotubes using an oxygen and carbon dual-free precursor. *RSC Adv.* **2018**, *8* (8), 3989–3995.

341. Terao, T.; Bando, Y.; Mitome, M.; Kurashima, K.; Zhi, C. Y.; Tang, C. C.; Golberg, D., Effective synthesis of surface-modified boron nitride nanotubes and related nanostructures and their hydrogen uptake. *Physica E.* **2008**, *40* (7), 2551–2555.

342. Lee, C. H.; Jieshang, W.; Vijaya, K. K.; Huang, J. Y.; Yap, Y. K., Effective growth of boron nitride nanotubes by thermal chemical vapor deposition. *Nanotechnology.* **2008**, *19* (45), 455605.

343. Lee, C. H.; Xie, M.; Kayastha, V.; Wang, J.; Yap, Y. K., Patterned growth of boron nitride nanotubes by catalytic chemical vapor deposition. *Chem. Mater.* **2010**, *22* (5), 1782–1787.

344. Huang, Y.; Lin, J.; Tang, C.; Bando, Y.; Zhi, C.; T., Z.; Dierre, B.; Sekiguchi, T.; Golberg, D., Bulk synthesis, growth mechanism and properties of highly pure ultrafine boron nitride nanotubes with diameters of sub-10 nm. *Nanotechnology.* **2011**, *22* (14), 145602.

345. Pakdel, A.; Zhi, C.; Bando, Y.; Nakayama, T.; Golberg, D., A comprehensive analysis of the CVD growth of boron nitride nanotubes. *Nanotechnology.* **2012**, *23* (21), 215601–215610.

346. Li, C.; Long, X.; E, S.; Zhang, Q.; Li, T.; Wu, J.; Yao, Y., Magnesium-induced preparation of boron nitride nanotubes and their application in thermal interface materials. *Nanoscale.* **2019**, *11* (24), 11457–11463.

347. Kim, M. J.; Chatterjee, S.; Kim, S. M.; Stach, E. A.; Bradley, M. G.; Pender, M. J.; Sneddon, L. G.; Maruyama, B., Double-walled boron nitride nanotubes grown by floating catalyst chemical vapor deposition. *Nano Lett.* **2008**, *8* (10), 3298–3302.

348. Chatterjee, S.; Kim, M. J.; Zakharov, D. N.; Kim, S. M.; Stach, E. A.; Maruyama, B.; Sneddon, L. G., Syntheses of boron nitride nanotubes from borazine and decaborane molecular precursors by catalytic chemical vapor deposition with a floating nickel catalyst. *Chem. Mater.* **2012**, *24* (15), 2872–2879.

349. Ma, R.; Bando, Y.; Sato, T., CVD synthesis of boron nitride nanotubes without metal catalysts. *Chem. Phys. Lett.* **2001**, *337* (1–3), 61–64.

350. Tay, R. Y.; Li, H.; Tsang, S. H.; Jing, L.; Tan, D.; Wei, M.; Teo, E. H. T., Facile synthesis of millimeter-scale vertically aligned boron nitride nanotube forests by template-assisted chemical vapor deposition. *Chem. Mater.* **2015**, *27* (20), 7156–7163.

351. McLean, B.; Eveleens, C. A.; Mitchell, I.; Webber, G. B.; Page, A. J., Catalytic CVD synthesis of boron nitride and carbon nanomaterials—Synergies between experiment and theory. *Phys. Chem. Chem. Phys.* **2017**, *19* (39), 26466–26494.

352. Wagner, R. S.; Ellis, W. C., Vapor-liquid-solid mechanism of single crystal growth. *Appl. Phys. Lett.* **1964**, *4* (5), 89–90.

353. Kumar, V.; Maity, P. C.; Lahiri, D.; Lahiri, I., Copper catalyzed growth of hexagonal boron nitride nanotubes on a tungsten substrate. *Cryst. Eng. Comm.* **2018**, *20* (19), 2713–2719.

354. Blase, X.; De Vita, A.; Charlier, J. C.; Car, R., Frustration effects and microscopic growth mechanisms for BN nanotubes. *Phys. Rev. Lett.* **1998**, *80* (8), 1666–1669.

355. Moon, W. H.; Hwang, H. J., Molecular-dynamics simulation of defect formation energy in boron nitride nanotubes. *Phys. Lett. A.* **2004**, *320* (5–6), 446–451.

356. Xu, H.; Ma, J.; Chen, X.; Hu, Z.; Huo, K.; Chen, Y., The electronic structures and formation mechanisms of the single-walled BN nanotube with small diameter. *J. Phys. Chem. B.* **2004**, *108* (13), 4024–4034.

357. Kuznetsov, V. L.; Mazov, I. N.; Delidovich, A. I.; Obraztsova, E. D.; Loiseau, A., Thermodynamic analysis of nucleation of boron nitride nanotubes on metal particles. *Phys. Status Solidi B.* **2007**, *244* (11), 4165–4169.

358. Wu, X.; Zeng, X. C., Adsorption of transition-metal atoms on boron nitride nanotube: A density-functional study. *J. Chem. Phys.* **2006**, *125* (4), 044711.

359. Zhao, J.-X.; Ding, Y.-H., Theoretical study of Ni adsorption on single-walled boron nitride nanotubes with intrinsic defects. *J. Phys. Chem. C.* **2008**, *112* (15), 5778–5783.

360. McLean, B.; Webber, G. B.; Page, A. J., Energy and charge transfer at the boron nitride nanotube—Catalyst growth interface. *J. Phys. Chem. C.* **2020**, *124* (21), 11662–11668.

361. Weismiller, M. R.; Duin, A. C. T. V.; Lee, J.; Yetter, R. A., ReaxFF reactive force field development and applications for molecular dynamics simulations of ammonia borane dehydrogenation and combustion. *J. Phys. Chem. A.* **2010**, *114* (17), 5485–5492.

362. McLean, B.; Webber, G. B.; Page, A. J., Boron nitride nucleation mechanism during chemical vapor deposition. *J. Phys. Chem. C.* **2018**, *122* (42), 24341–24349.

363. Krstic, P. S.; Han, L.; Irle, S.; Nakai, H., Simulations of the synthesis of boron-nitride nanostructures in a hot, high pressure gas volume. *Chem. Sci.* **2018**, *9* (15), 3803–3819.

364. Wang, Y.; Page, A. J.; Nishimoto, Y.; Qian, H.-J.; Morokuma, K.; Irle, S., Template effect in the competition between haeckelite and graphene growth on Ni(111): Quantum chemical molecular dynamics simulations. *J. Am. Chem. Soc.* **2011**, *133* (46), 18837–18842.

365. Krstic, P.; Han, L., Can hydrogen catalyze transitions between h-BN and c-BN in volume plasma? *J. Phys. Chem. C.* **2018**, *122* (1), 936–944.

366. McLean, B.; Webber, G. B.; Page, A. J., Boron nitride nanotube nucleation during Ni-catalyzed boron oxide chemical vapor deposition. *J. Phys. Chem. C.* **2019**, *123* (45), 27875–27883.

Chapter 4

Applications of BN Nanomaterials

4.1 Introduction

As established in Chapter 2, BN nanomaterials exhibit remarkable physical properties. Significant research has been performed to uncover the extent of these properties and how to tune them for potential applications. As a result, the synthesis of BN nanomaterials has been studied extensively with the vast parameter space of numerous techniques explored. Only with great knowledge of the properties of BN and the ability to routinely synthesize high-quality samples can a reliable and effective application be realized. Here, we discuss the applications of BN nanomaterials across numerous areas of industry and recent breakthroughs that show promise for their widespread use in the future.

4.2 Applications of c-BN

c-BN, perhaps the most widely utilized BN phase in today's industries, has long been used for its mechanical strength. c-BN has more attractive thermal properties compared to diamond, as well as additional benefits. Here, we discuss the use of c-BN in applications since its first synthesis over half a century ago and recent developments in perfecting its properties that may see it

Boron Nitride Nanomaterials: Properties, Fabrication, and Applications
Ben McLean and Alister J. Page

ISBN 978-981-4968-23-2 (Hardcover), 978-1-003-31448-6 (eBook)
www.jennystanford.com

overtake diamond as the hardest substance relevant for practical applications.

4.2.1 c-BN as an Abrasive

Shortly following its synthesis, c-BN was primarily used as a coating on cutting tools.[1, 2] Cutting ferrous materials with c-BN rather than diamond was attractive, since diamond reacts with iron at high temperatures (900–1,300 K).[3, 4] While advances in synthesis (Section 3.2) have led to high-quality c-BN film growth being relatively straightforward, the cost of these procedures on large scales is not yet feasible. As such, polycrystalline cubic boron nitride (PcBN) is used for mechanical applications such as cutting where it is combined with a binding material such as Al, Si, Mg, or similar. PcBN has proven effective for machining hard materials in multiple studies[5–8] at scalable rates for industrial applications. Early research showed that PcBN was completely inert during the machining of ferrous alloys.[9] It was later shown that there was some wear during the machining of hard ferrous materials, arising predominantly from chemical wear, attributed to BN diffusion[10] and the diffusion of trace elements such as Al, Si, and Mn in the working material.[11] Figure 4.1 shows three wear mechanisms *via* PcBN degradation during metal cutting that apply to both high (80–95%) c-BN and low (40–70%) c-BN content grades. Mechanical degradation of c-BN (Fig. 4.1a) has only been observed for high-speed milling applications.[12] Most commonly, wear in PcBN is attributed to either adhesive or chemical wear (Fig. 4.1b).[11, 13–17] The thermal stability of the binding material in industrial PcBN powders also influences the wear resistance of PcBN cutting tools.[14] Giménez *et al.*[18] combined experimental machining tests with a thermodynamic computational study to assess the role of chemical wear in the performance of both polycrystalline diamond (PcD) and PcBN between 700 °C and 1,300 °C. The results were in excellent agreement and indicated that PcBN materials were much more chemically stable than PcD, with the interaction of C with Fe and some of the binder species (W, Co) observed. Notably, the c-BN content of PcBN also matters when obtaining optimum performance and minimal wear.[19] Typically, low c-BN content grades show superior wear resistance for hard turning steels and case-hardened steels, while high c-BN content grades are better for hard milling of tool steels. Saketi *et al.*[20] performed effective hard milling of steels using the less commonly used high c-BN content tools indicating that controlling the cutting speed allows for continuous wear and a smooth tool surface due to tribochemical and adhesive wear, with little abrasive wear. They discuss three key forms of wear presented in Fig. 4.1. Overall, PcBN with low c-BN content provides the best performance in terms of tool life and surface finish. A low c-BN content tool produces a higher working temperature due to decreased thermal conductivity compared to high c-BN content and often results in a protective layer being formed, as well as softening the working material.[21] In addition, metallic binders are often used for high c-BN content tools while ceramics are used for low c-BN content tools, which bind stronger to the c-BN phase.[13] Monteiro *et al.*[22] reviewed the use of c-BN as a hard engineering material and collected the data shown in Table 4.1, comparing commercially available PcBN.

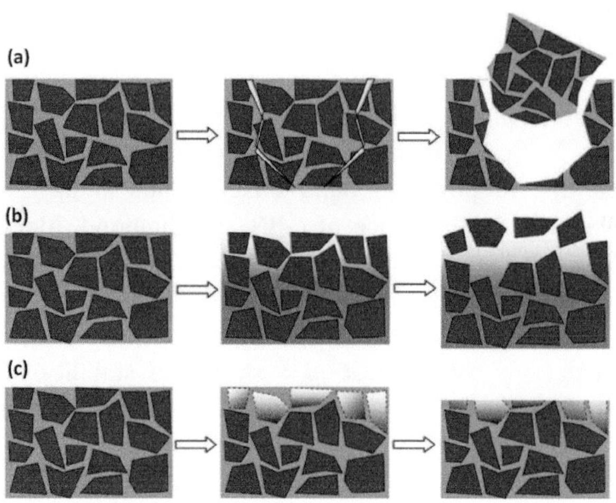

Figure 4.1 (a) Discrete mechanical degradation of PcBN during metal cutting. (b) Tribochemical wear of binder phase due to adhesion, abrasion, attrition, or diffusion leading to c-BN grain detachment. (c) Tribochemical wear of c-BN phase due to diffusion. Gray and black represent binder and c-BN phases, respectively.[20] Reprinted from Saketi, S. *et al.* Wear of a high cBN content PcBN cutting tool during hard milling of powder metallurgy cold work tool steels. *Wear* 2015, 332–333, 752–761, Copyright 2015, with permission from Elsevier.

Table 4.1 Mechanical properties of commercially available c-BN[22]

Brand	Density ($g\ cm^3$)	Knoop Hardness (GPa)	Young's Modulus (GPa)	Grain Size (μm)	Wear Intensity ($mm\ W^{-3}\ m^{-1}$)
BZN Compact®	3.48	35–45	740–870	1.0	11.8
BN 100®	4.25	40–45	830	0.3–2.0	12.6
BN 200®	4.25	30–35	814	0.5–1.5	12.2
Amborit®	3.37	28.5	680–720	0.1	12.1
Elbor®	3.31	32–38	660–750	0.5–1.2	12.4
Belbor®	3.42	39	730	0.2–1.5	10.6
Kiborit®	3.20	32–36	850–910	3.0–4.1	11.1

4.2.2 c-BN in Electronics

Being intrinsically electrically insulating, BN does not have obvious direct applications as a material in electronic devices. Instead, it is used in combination with other materials to yield new or improved electronic properties and/or performance. For instance, following the epitaxial formation of c-BN/diamond interfaces,[23-28] c-BN/diamond heterojunctions attracted research attention due to potential band offsets. Chen *et al.*[29] explored and elucidated the interface of c-BN and diamond, identifying a direct B-C bond at the interface, a potential electronic state for advanced electronic devices.[30] Zhao *et al.*[31] found this B-C bond configuration to be the most stable by first-principle calculations. DFT calculations by Shen *et al.*[32] show the stability of C-B-N-terminated diamond (111) surfaces that show promise for electron emission devices. c-BN films have also been realized as deep-UV photodetectors[33] and as a support for a patterned Mo/Au photodiode.[34] c-BN (and h-BN) has been used in photocatalysis on multiple occasions, as it suppresses electron-hole pair recombination and extends excitation wavelengths at only a few wt%. Zhou *et al.*[35] provide further discussion on the use of BN in composite photocatalytic materials.

4.3 Applications of h-BN

Among the 2D materials currently headlining materials science interests, h-BN is the premium electronic insulator for incorporation into low-dimensional heterostructures as a dielectric, compared to conducting TMDs and graphene.[36] h-BN additionally provides attractive thermal and mechanical properties that have seen its use in many other applications. We discuss each in the following subsections.

4.3.1 h-BN in Electronics

As mentioned in Chapter 2, Dean *et al.*[37] first constructed a graphene field effect transistor (FET) with an h-BN substrate, instead of SiO_2, and saw a remarkable enhancement of the carrier mobility (60,000 cm^2 V^{-1} s^{-1}). Soon thereafter, Wang *et al.*[38] developed an h-BN/graphene/h-BN FET for use in RF applications. Compared to a comparable Al_2O_3/graphene/SiO_2 device, the device containing h-BN performed significantly better with higher mobility and carrier velocity, as well as an improvement in the gain cut-off frequency, f_T, as shown in Fig. 4.2a. h-BN was subsequently widely used as a material replacing conventional dielectrics such as SiO_2 and Al_2O_3 with consistent improvements in mobilities,[39, 40] and low-frequency noise reduction.[41] Spectroscopic measurements by Xue *et al.*[42] showed that h-BN also reduces electron-hole fluctuations by two orders of magnitude compared to SiO_2, allowing for easier access to the Dirac point in graphene-based devices, verified by direct imaging of reduced fluctuations (Fig. 4.2b).[43]

In addition to RF devices, Gerbedoen *et al.*[44] demonstrated AlGaN/h-BN/GaN transistors as gate dielectrics with leakage currents lower than 20 pA mm^{-1} for microwave power devices. Gao *et al.*[45] showed that an h-BN/graphene-based electro-optic modulator outperformed analogous Si-based devices. Similarly, Shiue *et al.*[46] demonstrated the effectiveness of an h-BN/graphene photodetector. h-BN has also been integrated with MoS_2 into

FETs[47] and light-emitting diodes (LEDs) based on quantum wells.[48] Alternative graphene transistor architectures were explored; for example, vertical heterostructures were demonstrated by Britnell *et al.*,[49] utilizing vertical quantum tunneling and obtained high on-off ratios > 10^4. Wang *et al.*[50] encapsulated graphene with h-BN and made a 1D contact at the metallic graphene edge, enabling long-range ballistic transport and high electronic performance. Sandwiching graphene between h-BN in this way has been shown to enhance ballistic transport and protect graphene from the ambient atompshere.[51] Typically these heterostructure devices are made by mechanically transferring both h-BN and graphene into the device, though patterned growth[52, 53] and a co-segregation method[54] has also been achieved. Zhu *et al.*[55] produced thickness-sorted h-BN inks assembled into ultrathin layered dielectrics, which exhibited low leakage currents (3×10^{-9} A/cm^2 at 2 MV/cm) and high capacitances of 245 nF/cm^2). The use of h-BN in what is now referred to as van der Waals heterostructure has been reviewed by several authors in comprehensive reviews[56–64] and will undoubtedly be an integral part of next-generation electronic devices.[36] In addition, the band gap of graphene in contact with h-BN can be tuned by additional factors such as an applied electric field,[65] stacking,[66] and alignment.[67] The heterostructures, like the component materials themselves, can be doped to modify their properties.[68]

h-BN is also attractive for luminescent applications, such as LEDs and fluorescent screens,[69] as it is a deep UV emitter at room temperature[70] while effectively dissipating heat in the LED.[71] Watanabe *et al.*[72] fabricated a plane-emission device that emits far UV light from an h-BN fluorescent screen that could output 0.2 mW at 225 nm with low current consumption, using high-quality crystals that emit intense 215 nm luminescence at room temperature.[73] The potential for future photonic applications of h-BN is reviewed by Caldwell *et al.*[74] h-BN heterostructures have also been employed for sensing applications regarding pressure,[75–77] magnetic fields,[78–80] gases,[81, 82] biomolecules,[83] and pH.[84] For instance, Sajjad *et al.*[85] fabricated a prototype gas sensor for methane, showing high sensitivity and cyclability, while Lin *et al.*[86] showed h-BN nanosheets are also effective for sensing ethanol and NH_3.

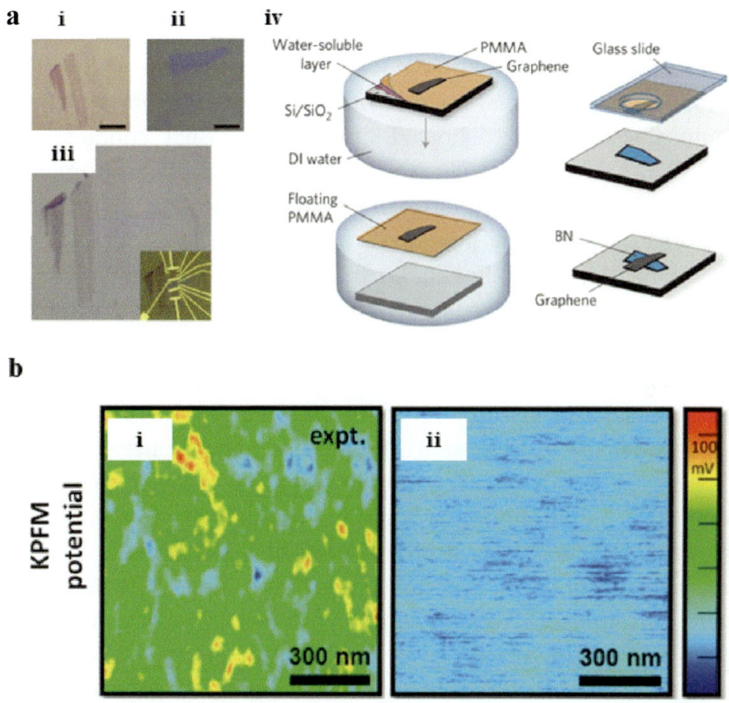

Figure 4.2 (a) (i) h-BN/graphene/h-BN device structure. (ii) Al_2O_3/graphene/SiO_2 device structure. (iii) Comparison of the devices in (a) and (b) in terms of gain and frequency, showing maximum gain f_T.[38] Reprinted by permission from Springer Nature Customer Service Center GmbH: Springer Nature Dean, C. R. *et al.* Boron nitride substrates for high-quality graphene electronics. *Nat. Nanotech.* 2010, 5, 722. Copyright 2010. (b) Topography and relative potential for (i) SiO_2/Si and (ii) exfoliated h-BN on SiO_2.[43] Reprinted with permission from Burson, K. M. *et al.* Direct imaging of charged impurity density in common graphene substrates. *Nano Lett.* 2013, 13 (8), 3576–3580. Copyright 2013 American Chemical Society.

4.3.2 h-BN in Thermal Applications

Due to the excellent thermal conductivities and electrically insulating properties of h-BN, there has been significant research into fabricating nanocomposite materials with h-BN and polymers. To enhance bonding with the polymer to form interconnected matrices, h-BN is often functionalized for use in these composites,[87, 88] as the properties such as hydrophobicity of h-BN can be tuned to engineer the interaction in the desired composite material.[89] Li and Hsu[90, 91] developed h-BN/polyimide composite films that showed enhanced thermal conductivities (1.2 W m^{-1} K^{-1}) compared to the base polymer; these conductivities increased further with increasing h-BN particle size. Shoji *et al.*[92] dispersed h-BN with polyimide in its liquid crystalline phase, observing that cross-linked LC polyimide maintained alignment even with high h-BN content. This yielded a stronger film that still exhibited thermal diffusivity higher than amorphous polyimide itself. Yoshihara *et al.*[93] also used LC-phase polymers (polyesters) as a medium for the dispersion of h-BN and saw that a perpendicular orientation of molecular chains between h-BN sheets can act as heat paths to enhance thermal properties. A 17-fold enhancement to thermal conductivity was observed by Wang *et al.*[94] for h-BN/polymethyl methacrylate (PMMA) composites. Enhanced thermal conductivities up to 5.24 W m^{-1} K^{-1} were observed by Wang *et al.*[95] for h-BN/epoxy composites, a 1,600% improvement over the pristine epoxy. Other h-BN/epoxy[96, 97] and h-BN/polymer[98–100] composites saw enhanced thermal conductivities compared to the respective base polymers, and were tunable based on the orientation[101] and functionalization[102] of the h-BN within the matrix. Song *et al.*[103] showed that, *via* mechanical stretching, a composite of h-BN and polyvinyl alcohol (PVA), exhibit high thermal conductivities (> 30 W m^{-1} K^{-1}) induced by strain-induced alignments of nanosheets within the composite structure. The inclusion of h-BN into mineral oil (MO)[104] and cellulose-based papers[105] saw high thermal conductivities in materials promising as replacements for traditional industry standards (Fig. 4.3). In addition, due to their low thermal diffusivity and effusivity, h-BN coatings applied to wood demonstrated oxidation and fire resistance up to 900 °C.[106] Ambekar *et al.*[107] fabricated h-BN-coated cotton fabrics that were flame retardant, hydrophobic, and revealed self-extinguishing properties, with no vertical or lateral flame movement on the material.

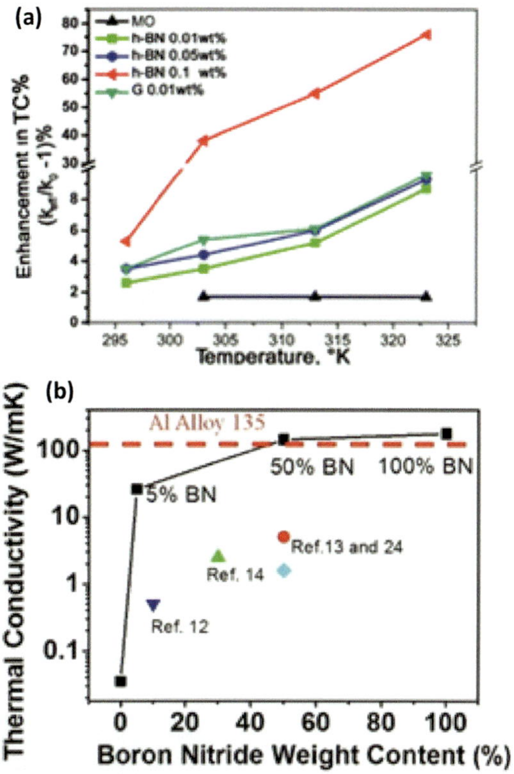

Figure 4.3 (a) Temperature-dependent effective thermal conductivity enhancement of various nanofluids with different h-BN and graphene (G) content in MO.[104] Reprinted with permission from Taha-Tijerina, J. *et al.* Electrically insulating thermal nano-oils using 2D fillers. *ACS Nano* 2012, 6 (2), 1214–1220. Copyright 2012 American Chemical Society. ((b) Thermal conductivity of the h-BN/cellulose composite as a function of h-BN weight content, reaching 145.7 W m^{-1} K^{-1} at 50% h-BN.[105] Reprinted with permission from Zhu, H. *et al.* Highly thermally conductive papers with percolative layered boron nitride nanosheets. *ACS Nano* 2014, 8 (4), 3606–3613. Copyright 2014 American Chemical Society.

4.3.3 h-BN in Mechanical Applications

As h-BN exhibits excellent mechanical properties, such as high Young's modulus and fracture strength (Chapter 2), it has been utilized in polymeric composites[108] as a means of mechanical reinforcement in addition to thermal enhancement as discussed above. Jan *et al.*[109] found that placing the h-BN/polymer composites under strain maximized their mechanical properties compared to the pure polymer. Zhi *et al.*[110] fabricated composite films with PMMA and showed that embedding h-BN (only 0.3 wt%) increased their Young's modulus (by 22%) and fracture strength (by 11%). On the other hand, composites including BNNTs only increased Young's modulus due to defect concentration. Sainsbury *et al.*[111] observed significant enhancement of h-BN/PVA composite (0.1 wt% h-BN) mechanical properties when the h-BN was hydroxylated as per Fig. 4.4. h-BN nanosheets have also been used to reinforce glass matrix composites, with Saggar *et al.*[112] observing a 45% increase in fracture and flexure strength with only 5% h-BN loading into borosilicate glass.

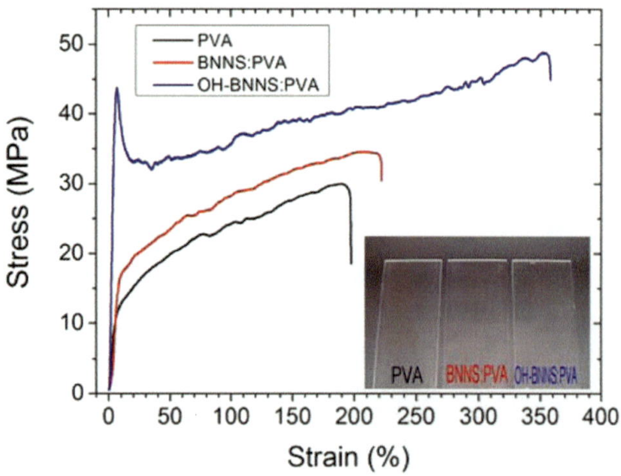

Figure 4.4 Representative stress-strain curves for pure PVA film and films of h-BN/PVA (BNNS/PVA) and hydroxylated h-BN/PVA (OH-BNNS: PVA) with transparent samples on glass microscope slides as inset.[111] Reprinted with permission from Sainsbury, T. *et al.* Oxygen radical functionalization of boron nitride nanosheets. *J. Am. Chem. Soc.* 2012, 134 (45), 18758–18771. Copyright 2012 American Chemical Society.

Pure h-BN layers have also been utilized as protective films. For instance, Liu *et al.*[113] utilized h-BN as oxidation-resistant coatings for nickel and graphene up to 1,100 °C in oxidizing atmospheres while maintaining their mechanical integrity. Coating metals such as Cu with h-BN have proven effective for corrosion protection in saltwater,[114] electrolyte solutions,[115] and in sulfurous[116] and biological[117] environments. Many epoxy coatings have thus demonstrated increased corrosion resistance when containing h-BN.[118–120] Notably, however, Khan *et al.*[114] demonstrated that the quality of h-BN drastically influences the performance in this respect, with grain boundaries and cracks promoting the inclusion of H and O radicals that enable corrosion of the substrate.

4.4 Applications of BNNTs

As discussed in Section 2.2.3, BNNTs exhibit remarkable mechanical, thermal, and electrical applications irrespective of chirality, unlike CNTs. In a sense, this is a blessing and a curse; as they are intrinsically dielectrics, their electronic properties cannot be 'tuned' like their carbon-based cousins, but this means there is no need to strive for diameter or chirality specific synthesis (which makes their production potentially easier and therefore cheaper). Nevertheless, despite the promise BNNTs hold for future application, there is currently a lack of means for mass production, compared to their CNT analogs, although recent methodologies are emerging for purification and quality assessment on larger scales.[121] Here, we discuss the use of BNNTs in composite materials and devices, and the resulting enhancement of potentially useful materials.

4.4.1 BNNTs in Mechanical Applications

Zhi *et al.*[122] first fabricated BNNT/polymer composite films with polystyrene (PS) and observed a 21% enhancement in Young's modulus with only 1 wt% BNNT content, in addition to an increase in the oxidative stability and thermal conductivity. Zhi *et al.*[123] also fabricated polymeric composites with BNNTs and PMMA, PS, poly(vinyl butyral) (PVB), and poly(ethylene vinyl alcohol) (PEVA). These composites also exhibited an improved thermal conductivity, electric breakdown field, and Vickers microhardness. These studies paved the way for BNNTs to be utilized in a range of composites and in their pristine forms. Zhi *et al.*[124] expanded their study of BNNT/polymer composites by functionalizing BNNTs with –OH and –NH groups. The hydroxylated BNNTs were able to bind more effectively with the polymer and further enhance the mechanical performance (Young's modulus 2.75 GPa) compared to the base polymer (2.25 GPa) and the composite containing un-functionalized BNNTs (2.5 GPa). Similar functionalization and formation of BNNT/polymer composites have seen improvements in mechanical performance.[125–128] Figure 4.5 presents the macroscopic appearance and microscopic structure of BNNT

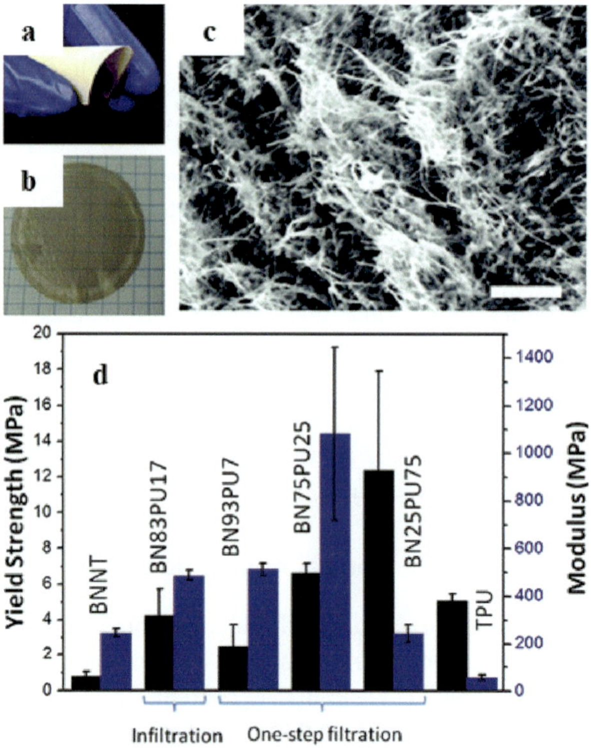

Figure 4.5 Photographs of (a) 75 μm and (b) 35 μm thick composites of BNNT and thermopolyurethane (TPU). (c) SEM cross-section of a BNNT/resin buckypaper (BN83PU17). (d) Yield strength and Young's modulus of neat TPU, BNNT buckypaper, and BNNT/TPU composites.[128] Kim, K. S. *et al.* Polymer nanocomposites from free-standing, macroscopic boron nitride nanotube assemblies. *RSC Adv.* 2015, 5 (51), 41186–41192. Published by The Royal Society of Chemistry.

composites produced by Kim *et al.*[128] as well as their yield strengths and Young's modulus. Yamaguchi *et al.*[129] showed that BNNTs can be integrated into an Al matrix to more than double the matrix's tensile strength. DFT calculations by Azizi *et al.*[130] showed that these Al-BNNT composites also have the potential for methane sensing and storage. Wang *et al.*[131] recently reported a highly compressive BNNT/reduced graphene oxide (rGO) aerogel. This material displayed high compression resilience, enhanced thermal conductivity, and the ability to recover effectively after large strain.

Yarns or fibers are a major mechanical and electrical application for CNTs, and can now be routinely synthesized with remarkable properties.[132–135] Similarly, BNNT-based yarns synthesized *via* laser vaporization have emerged as versatile and strong materials.[136] Fabrication of BNNT yarns has now reached pilot-scale production rates of 20 g h^{-1} by Kim *et al.*[137] and 35 g h^{-1} by Fathalizadeh *et al.*[138] using plasma techniques.

4.4.2 BNNTs in Thermal Applications

Huang *et al.*[139] functionalized BNNTs with polyhedral oligosilsesquioxane (POSS) and fabricated epoxy composites that showed an increase in thermal conductivity of 1,360% with 30% BNNT content. BNNT/polymer composites also exhibit a large increase in thermal conductivity at low BNNT content.[140–144] BNNT/cellulose composites fabricated by Zeng *et al.*[145] exhibit higher thermal conductivity than BNNT/polymer composites, and demonstrated their utility as printed circuit boards within a flexible cooling electronic device (Fig. 4.6). Similar composite papers formed using BNNTs and polyimide exhibit unique flame and structural deformation resistance. [128] Pristine and hydroxylated BNNTs have also been incorporated in water-based thermally conductive nanofluids water containing a polydiallyldimethylammonium chloride (PDDA) stabilizer.[146] The BNNT/polymer composites were shown to be more effective than traditional species such as CuO, MgO, and Al_2O_3. Due to impressive thermal dissipation properties, BNNTs were utilized by Seo *et al.*[147] in GaN LEDs, which exhibited enhanced luminescence intensity and lower surface temperature.

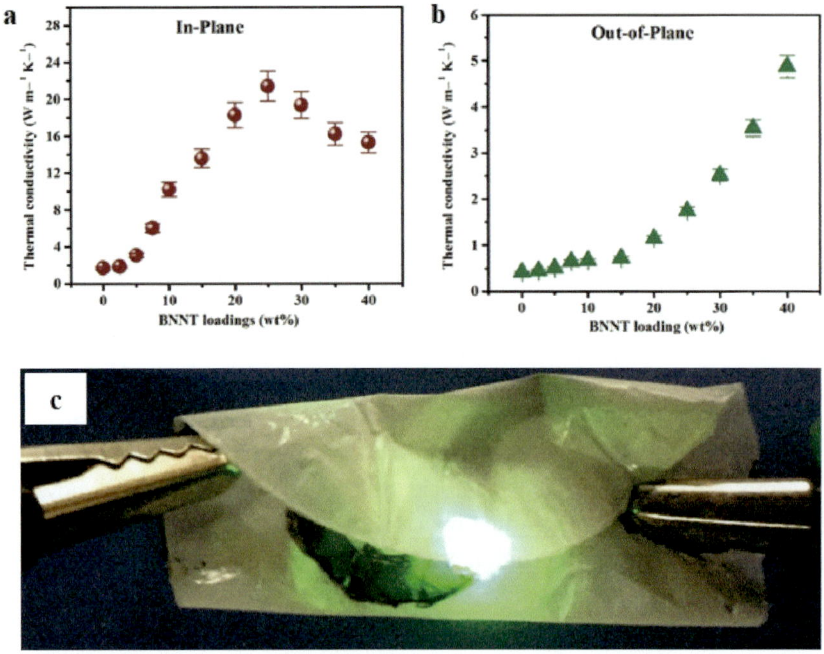

Figure 4.6 Thermal conductivity of the BNNT/cellulose composite (a) in-plane and (b) out-of-plane. (c) Optical image of the working electronic devices by using a BNNT/cellulose composite as a flexible, printed circuit board.[145] Reprinted with permission from Zeng, X. *et al.* A combination of boron nitride nanotubes and cellulose nanofibers for the preparation of a nanocomposite with high thermal conductivity. *ACS Nano* 2017, 11 (5), 5167–5178. Copyright 2017 American Chemical Society.

4.4.3 BNNTs in Electronics

While exhibiting similar electrical properties to 2D h-BN, 1D BNNTs have not made a significant impact in electronic applications, unlike their 2D counterpart. Nevertheless, there are a number of avenues regarding spintronic applications where BNNTs may find applications. Many of these revolve around doped-BNNTs. For instance, BNNTs have been doped with B,[148] C,[149–152] F,[153, 154] Si,[155] Ge,[156] and transition metals,[157, 158] to name a few, with numerous heteroatoms, again inducing magnetism.[159] Hao *et al.*[160] demonstrated that pure BNNTs can realize their intrinsic magnetism *via* their open ends, depending on the open edge exposed (and hence the chirality of the BNNT). B-rich edges of open BNNTs exhibit high spin polarizations, which in principle enable their use in spin-polarized electron field emitters and STM tips. C ad-atoms also induce a large net magnetic moment, as calculated by Li *et al.*,[161] independent of the tube diameter and bonding site. In contrast, adsorbing F atoms to the BNNT induces diameter-dependent ferromagnetism (magnetic moment significantly increases with decreasing tube diameter).[153] Each instance of induced magnetism promises BNNTs may be useful in future spintronic devices., particularly if the material is both semiconducting and ferromagnetic; making it a stable magnetic semiconductor, as in the case of the hybrid Ni-BNNTs synthesized by Lenin *et al.*[162] *via* CVD.

The progress of h-BN as an important material for electronics generally stems from its inclusion in heterostructures, typically (although not always) in conjunction with its structural analog, graphene. For BNNTs, the idea of a heterostructure with CNTs was predicted to display unique electronic properties (Fig. 4.7), whether it be as a coaxial structure[163, 164] or a heterojunction.[165] Heterojunctions between CNTs and BNNTs show promise for nano rectifiers with attractive electrical[165, 166] and thermal[167] transport properties across the interface. CNT-BNNT-CNT nanostructures have also been predicted to have gas-sensing properties with high selectivity toward NO_2, O_2, and H_2O gases.[168] On the other hand, a landmark synthesis by Xiang *et al.*[169] achieved 1D, monocrystalline coaxial van der Waals heterostructures consisting of an inner CNT and outer BNNT (CNT@BNNT) as well as a CNT@BNNT@MoS_2 structure. The BNNT successfully thermally protects the inner CNT without sacrificing the superior electrical properties of the CNT, showing good performance in a FET. This recent discovery opens a wide range of possibilities for BNNTs much like 2D heterostructures have for h-BN.

A related strategy is to combine CNTs and BNNTs non-coaxially. For instance, Kim *et al.*,[128] fabricated a "sandwich cookie buckypaper" consisting of alternating layers of CNTs and BNNTs. Kang *et al.*[170] constructed a similar CNT/BNNT electroactive actuator with impressive piezoelectric and electrostrictive coefficients compared to a BNNT/polyimide composite (Fig. 4.8). The attractive optical properties and neutron shielding exhibited by BNNTs suggested that these actuators may be effective for space exploration and similar extreme environments.

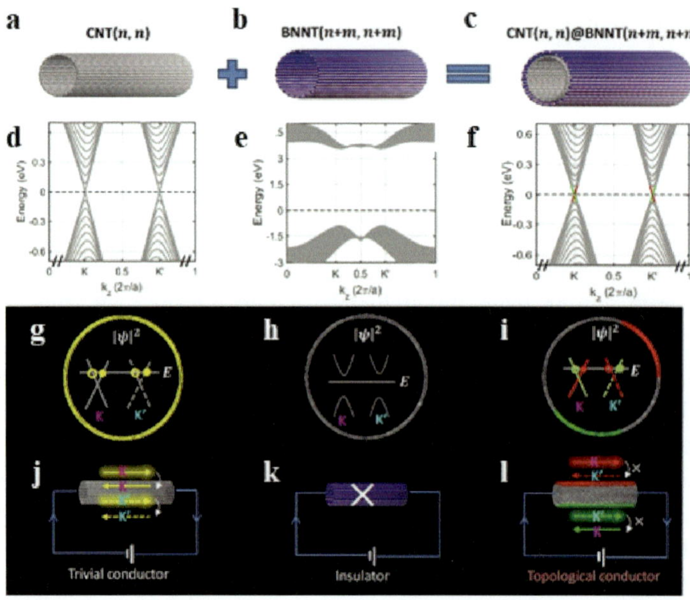

Figure 4.7 Geometries of armchair (a) CNT, (b) BNNT, and (c) CNT@BNNT. Calculated band structures demonstrating the nature of the band gap for (d) CNT, (e) BNNT, and (f) CNT@BNNT. Spatial distribution of modular squared wave functions on the tube circumference for (g) CNT, (h) BNNT, and (i) CNT@BNNT. Schematics of charge transport in (j) CNT, (k) BNNT, and (l) CNT@BNNT.[164] Reprinted with permission from Hu, C. *et al.* Theoretical design of topological heteronanotubes. *Nano Lett.* 2019, 19 (6), 4146–4150. Copyright 2019 American Chemical Society.

4.4.4 BNNTs in Biological Applications

In addition to being mechanically strong and chemically resistant, BNNTs have also shown to be non-cytotoxic,[171–173] opening avenues for their potential use as additives in bioactive ceramics[174] and in drug delivery systems.[175]

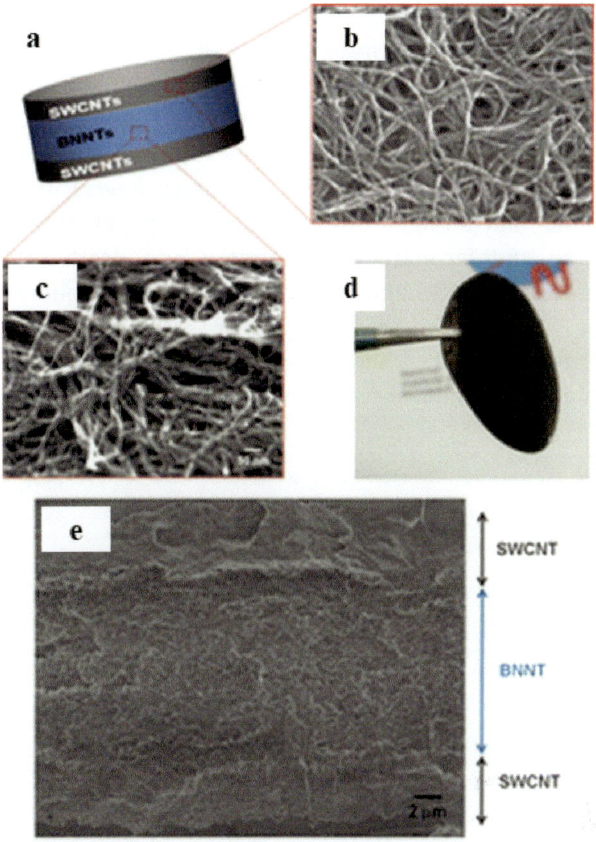

Figure 4.8 (a) Schematic structure of the CNT/BNNT/CNT actuator fabricated by Kang *et al.* SEM images of the (b) top and bottom CNT layers and (c) BNNT core layer. (d) Photograph of the free-standing prototype. (e) SEM image of the actuator cross-section.[170] Reprinted with permission from Kang, J. H. *et al.* Multifunctional electroactive nanocomposites based on piezoelectric boron nitride nanotubes. *ACS Nano* 2015, 9 (12), 11942–11950. Copyright 2015 American Chemical Society.

Chen *et al.*[176] demonstrated the non-cytotoxic nature of BNNTs, showing cells underwent no change with prolonged exposure to BNNTs, unlike CNTs. By functionalization, BNNTs were also able to dissolve in an aqueous environment and interface selectively with chosen proteins; aqueous solubility is a well-known impediment for using CNTs in similar contexts. Using single-stranded deoxyribonucleic acid (DNA), BNNTs were tested as drug-delivery agents and proved effective at delivering DNA oligomers to the interior of cells without any cellular toxicity. However, Horváth *et al.*[177] highlighted the importance of a dispersing agent to stabilize the BNNT suspension in this respect (without which, BNNTs were reported to be cytotoxic in the same cell line at a dose 5,000 times lower). In addition, they found that straight BNNTs could easily penetrate the cell membrane and facilitate cell alteration and death (Fig. 4.9a). This *in vitro* study and several others that followed[178,179], however, are in contrast to other studies demonstrating non-cytotoxic BNNTs (including functionalized BNNTs)[180–182] that did not induce cell or DNA damage (Fig. 4.9b).

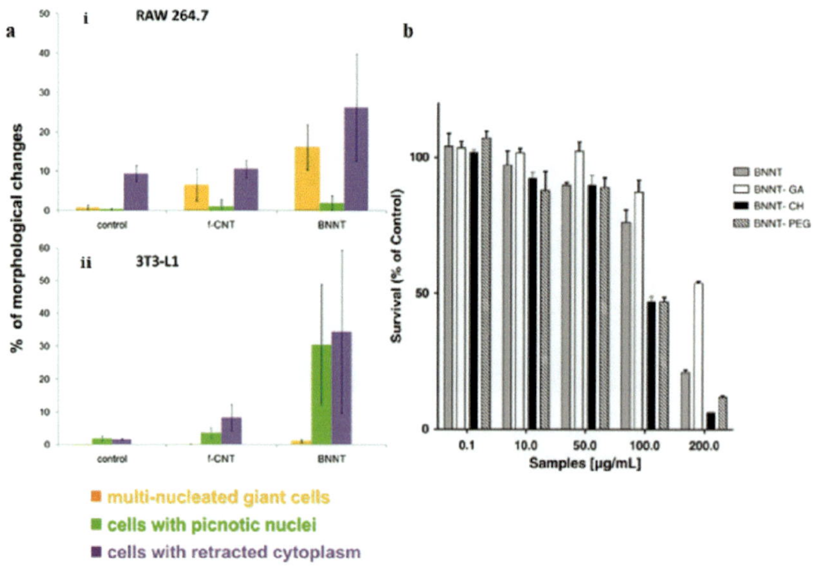

Figure 4.9 (a) Quantification of morphological changes for (i) RAW 264.7 macrophages and (ii) 3T3-L1 fibroblasts determined morphological changes.[177] Reprinted with permission from Horváth, L. *et al. in vitro* investigation of the cellular toxicity of boron nitride nanotubes. *ACS Nano* 2011, 5 (5), 3800–3810. Copyright 2011 American Chemical Society. (b) Cytotoxicity of BNNTs and functionalized BNNTs on MRC-5 cell line.[182] Reprinted from Ferreira, T. H. *et al.* Boron nitride nanotubes coated with organic hydrophilic agents: Stability and cytocompatibility studies. *Mater. Sci. Eng. C* 2013, 33 (8), 4616–4623, Copyright 2013, with permission from Elsevier.

Biocompatibility of BNNTs more generally was studied by Ciofani *et al.*,[183, 184] who demonstrated that they have no adverse effects or impairments in blood, liver, and kidney functions. Similarly, Soares *et al.*[185] reported in vivo accumulation of BNNTs in the liver, spleen, and gut followed by elimination. Li *et al.*[186] also demonstrated BNNT/silica hybrids that were stable and effective at delivering doxorubicin, a common chemotherapy agent. Following *in vitro* tests demonstrating feasibility for influencing the uptake of BNNTs by an external magnetic source,[187] Li *et al.*[188] fabricated luminescent-magnetic BNNTs encapsulated in $NaGdF_4$:Cu composites that allowed for effective intracellular drug delivery of chemotherapy agents in the presence of an external magnetic field. DFT calculations have also shown Al-doped BNNTs to be preferentially bound to common chemotherapy agents cis- and neda-platin.[189]

4.4.5 BNNTs in Sensing Applications

SW and DWBNNTs were utilized by Cho *et al.*[190] as an adsorbent/sensor for purifying water of methylene blue with an efficiency of 94%. BNNTs were added to a solution of methylene blue and UV-Vis absorption were spectra taken over time showed that the methylene blue was captured from the solution by the BNNTs. Aside from selective chemical sensing, BNNTs have been utilized for pH sensing when functionalized by fluorescin and Ag nanoparticles,[191] and for humidity when decorated with Au atoms.[192] A BNNT-based contact electrification-assisted piezoelectric nanogenerator (0.5 % BNNTs, 0.1 % CNTs) was employed by Kim *et al.*[193] as a kinematic sensor to detect robotic motion. There is great potential for BNNTs to be used as nanosensors in solution, as they permeate water effectively, while conducting it.[194]

References

1. Corrigan, F. R. S., D. E. Process for making cubic boron nitride from coated hexagonal boron nitride, and abrasive particles and articles made therefrom. **1989**.

2. Wentorf, R. H.; De Lai, A. Cubic boron nitride compact and method for its production. **1964**.

3. Shimada, S.; Tanaka, H.; Higuchi, M.; Yamaguchi, T.; Honda, S.; Obata, K., Thermo-chemical wear mechanism of diamond tool in machining of ferrous metals. *CIRP Annals.* **2004**, *53* (1), 57–60.

4. Tanaka, T.; Ikawa, N., Thermal aspect of wear of diamond abrasive grain in grinding mild steel. *J. Jpn. Soc. Prec. Eng.* **1973**, *39* (461), 608–612.

5. Jia, Y.; Ye, Z.; Wang, H.; Jing, H., Techanics study on PcBN cutting tool dry turning chilled cast iron. In *Adv. Mater. Res.* **2012**, vol. 418–420, 1342–1345.

6. Keunecke, M.; Wiemann, E.; Weigel, K.; Park, S. T.; Bewilogua, K., Thick c-BN coatings—Preparation, properties and application tests. *Thin Solid Films.* **2006**, *515* (3), 967–972.

7. Mahfoudi, F.; List, G.; Molinari, A.; Moufki, A.; Boulanouar, L., High speed turning for hard material with PcBN inserts: Tool wear analysis. *Int. J. Mach. Mach. Mater.* **2008**, *3* (1–2), 62–79.

8. Shao, F.; Wang, Y.; Wang, Y.; Qun, C., Wear of PcBN tool when cutting materials difficult-to-cut based on thermodynamics entropy. *Metall. Min. Ind.* **2015**, *7* (4), 381–389.

9. Narutaki, N.; Yamane, Y., Tool wear and cutting temperature of cbn tools in machining of hardened steels. *Annals of the ICRP.* **1979**, *28* (1), 23–28.

10. Zimmermann, M.; Lahres, M.; Viens, D. V.; Laube, B. L., Investigations of the wear of cubic boron nitride cutting tools using Auger electron spectroscopy and X-ray analysis by EPMA. *Wear.* **1997**, *209* (1), 241–246.

11. Barry, J.; Byrne, G., Cutting tool wear in the machining of hardened steels: Part II: cubic boron nitride cutting tool wear. *Wear.* **2001**, *247* (2), 152–160.

12. Liu, Z. Q.; Ai, X.; Zhang, H.; Wang, Z. T.; Wan, Y., Wear patterns and mechanisms of cutting tools in high-speed face milling. *J. Mater. Process. Technol.* **2002**, *129* (1), 222–226.

13. Klimenko, S. A.; Mukovoz, Y. A.; Lyashko, V. A.; Vashchenko, A. N.; Ogorodnik, V. V., On the wear mechanism of cubic boron nitride base cutting tools. *Wear.* **1992**, *157* (1), 1–7.

14. König, W.; Neises, A., Wear mechanisms of ultrahard, non-metallic cutting materials. *Wear.* **1993**, *162–164*, 12–21.

15. Angseryd, J.; Coronel, E.; Elfwing, M.; Olsson, E.; Andrén, H. O., The microstructure of the affected zone of a worn PcBN cutting tool characterised with SEM and TEM. *Wear.* **2009,** *267* (5), 1031–1040.

16. Lahiff, C.; Gordon, S.; Phelan, P., PcBN tool wear modes and mechanisms in finish hard turning. *Robot. Comput. Integr. Manuf.* **2007,** *23* (6), 638–644.

17. Huang, Y.; Chou, Y. K.; Liang, S. Y., CBN tool wear in hard turning: a survey on research progresses. *Int. J. Adv. Manuf. Tech.* **2007,** *35* (5), 443–453.

18. Giménez, S.; Van der Biest, O.; Vleugels, J., The role of chemical wear in machining iron based materials by PcD and PcBN super-hard tool materials. *Diamond Relat. Mater.* **2007,** *16* (3), 435–445.

19. Cook, M. W.; Bossom, P. K., Trends and recent developments in the material manufacture and cutting tool application of polycrystalline diamond and polycrystalline cubic boron nitride. *Int. J. Refract. Met. Hard Mater.* **2000,** *18* (2), 147–152.

20. Saketi, S.; Sveen, S.; Gunnarsson, S.; M'Saoubi, R.; Olsson, M., Wear of a high cBN content PcBN cutting tool during hard milling of powder metallurgy cold work tool steels. *Wear.* **2015,** *332–333,* 752–761.

21. Hooper, R. M.; Parry, A.; Shakib, J. I.; Brookes, C. A., Mechanical properties, microstructure and wear of DBC50. *Ind. Diamond Rev.* **1989,** *49* (533), 170, 172–173.

22. Monteiro, S. N.; Skury, A. L. D.; de Azevedo, M. G.; Bobrovnitchii, G. S., Cubic boron nitride competing with diamond as a superhard engineering material – An overview. *J. Mater. Res. Technol.* **2013,** *2* (1), 68–74.

23. Koizumi, S.; Murakami, T.; Inuzuka, T.; Suzuki, K., Epitaxial growth of diamond thin films on cubic boron nitride {111} surfaces by DC plasma chemical vapor deposition. *Appl. Phys. Lett.* **1990,** *57* (6), 563–565.

24. Yoshikawa, M.; Ishida, H.; Ishitani, A.; Koizumi, S.; Inuzuka, T., Study of crystallographic orientations in the diamond film on the (100) surface of cubic boron nitride using a Raman microprobe. *Appl. Phys. Lett.* **1991,** *58* (13), 1387–1388.

25. Zhang, X. W.; Boyen, H. G.; Deyneka, N.; Ziemann, P.; Banhart, F.; Schreck, M., Epitaxy of cubic boron nitride on (001)-oriented diamond. *Nat. Mater.* **2003,** *2* (5), 312–315.

26. Taniguchi, T.; Yamaoka, S., Heteroepitaxial growth of cubic boron nitride single crystal on diamond seed under high pressure. *MRS Proc.* **1997,** *472,* 379.

27. Hirama, K.; Taniyasu, Y.; Yamamoto, H.; Kumakura, K., Structural analysis of cubic boron nitride (111) films heteroepitaxially grown on diamond (111) substrates. *J. Appl. Phys.* **2019**, *125* (11), 115303.

28. Zhang, W.; Bello, I.; Lifshitz, Y.; Chan, K. M.; Meng, X.; Wu, Y.; Chan, C. Y.; Lee, S. T., Epitaxy on diamond by chemical vapor deposition: A route to high-quality cubic boron nitride for electronic applications. *Adv. Mater.* **2004**, *16* (16), 1405–1408.

29. Chen, C.; Wang, Z.; Kato, T.; Shibata, N.; Taniguchi, T.; Ikuhara, Y., Misfit accommodation mechanism at the heterointerface between diamond and cubic boron nitride. *Nat. Commun.* **2015**, *6* (1), 6327.

30. Lin, Z.; Peng, X.; Huang, C.; Fu, T.; Wang, Z., Atomic structure, electronic properties and generalized stacking fault energy of diamond/c-BN multilayer. *RSC Adv.* **2017**, *7* (47), 29599–29605.

31. Zhao, D.; Gao, W.; Li, Y.; Zhang, Y.; Yin, H., The electronic properties and band-gap discontinuities at the cubic boron nitride/diamond hetero-interface. *RSC Adv.* **2019**, *9* (15), 8435–8443.

32. Shen, W.; Shen, S.; Liu, S.; Li, H.; Gan, Z.; Zhang, Q., Monolayer cubic boron nitride terminated diamond (111) surfaces for quantum sensing and electron emission applications. *ACS Appl. Mater. Int.* **2020**, *12* (29), 33336–33345.

33. Soltani, A.; Barkad, H. A.; Mattalah, M.; Benbakhti, B.; De Jaeger, J. C.; Chong, Y. M.; Zou, Y. S.; Zhang, W. J.; Lee, S. T.; BenMoussa, A.; Giordanengo, B.; Hochedez, J. F., 193nm deep-ultraviolet solar-blind cubic boron nitride based photodetectors. *Appl. Phys. Lett.* **2008**, *92* (5), 053501.

34. BenMoussa, A.; Soltani, A.; Schühle, U.; Haenen, K.; Chong, Y. M.; Zhang, W. J.; Dahal, R.; Lin, J. Y.; Jiang, H. X.; Barkad, H. A.; BenMoussa, B.; Bolsee, D.; Hermans, C.; Kroth, U.; Laubis, C.; Mortet, V.; De Jaeger, J. C.; Giordanengo, B.; Richter, M.; Scholze, F.; Hochedez, J. F., Recent developments of wide-bandgap semiconductor based UV sensors. *Diamond Relat. Mater.* **2009**, *18* (5), 860–864.

35. Zhou, C.; Lai, C.; Zhang, C.; Zeng, G.; Huang, D.; Cheng, M.; Hu, L.; Xiong, W.; Chen, M.; Wang, J.; Yang, Y.; Jiang, L., Semiconductor/boron nitride composites: Synthesis, properties, and photocatalysis applications. *Appl. Catal. B* **2018**, *238*, 6–18.

36. Lim, H.; Yoon, S. I.; Kim, G.; Jang, A. R.; Shin, H. S., Stacking of two-dimensional materials in lateral and vertical directions. *Chem. Mater.* **2014**, *26* (17), 4891–4903.

37. Dean, C. R.; Young, A. F.; Meric, I.; Lee, C.; Wang, L.; Sorgenfrei, S.; Watanabe, K.; Taniguchi, T.; Kim, P.; Shepard, K. L.; Hone, J., Boron

nitride substrates for high-quality graphene electronics. *Nat. Nanotech.* **2010**, *5*, 722.

38. Wang, H.; Taychatanapat, T.; Hsu, A.; Watanabe, K.; Taniguchi, T.; Jarillo-Herrero, P.; Palacios, T., BN/graphene/BN transistors for RF applications. *IEEE Electron Device Lett.* **2011**, *32* (9), 1209–1211.
39. Bresnehan, M. S.; Hollander, M. J.; Wetherington, M.; Wang, K.; Miyagi, T.; Pastir, G.; Snyder, D. W.; Gengler, J. J.; Voevodin, A. A.; Mitchel, W. C.; Robinson, J. A., Prospects of direct growth boron nitride films as substrates for graphene electronics. *J. Mater. Res.* **2014**, *29* (3), 459–471.
40. Wang, L.; Wu, B.; Chen, J.; Liu, H.; Hu, P.; Liu, Y., Monolayer hexagonal boron nitride films with large domain size and clean interface for enhancing the mobility of graphene-based field-effect transistors. *Adv. Mater.* **2014**, *26* (10), 1559–1564.
41. Stolyarov, M. A.; Liu, G.; Rumyantsev, S. L.; Shur, M.; Balandin, A. A., Suppression of 1/f noise in near-ballistic h-BN-graphene-h-BN heterostructure field-effect transistors. *Appl. Phys. Lett.* **2015**, *107* (2), 023106.
42. Xue, J.; Sanchez-Yamagishi, J.; Bulmash, D.; Jacquod, P.; Deshpande, A.; Watanabe, K.; Taniguchi, T.; Jarillo-Herrero, P.; LeRoy, B. J., Scanning tunnelling microscopy and spectroscopy of ultra-flat graphene on hexagonal boron nitride. *Nat. Mater.* **2011**, *10* (4), 282–285.
43. Burson, K. M.; Cullen, W. G.; Adam, S.; Dean, C. R.; Watanabe, K.; Taniguchi, T.; Kim, P.; Fuhrer, M. S., Direct imaging of charged impurity density in common graphene substrates. *Nano Lett.* **2013**, *13* (8), 3576–3580.
44. Gerbedoen, J. C.; Soltani, A.; Mattalah, M.; Moreau, M.; Thevenin, P.; De Jaeger, J. C., AlGaN/GaN MISHEMT with h-BN as gate dielectric. *Diamond Relat. Mater.* **2009**, *18* (5), 1039–1042.
45. Gao, Y.; Shiue, R.-J.; Gan, X.; Li, L.; Peng, C.; Meric, I.; Wang, L.; Szep, A.; Walker, D.; Hone, J.; Englund, D., High-speed electro-optic modulator integrated with graphene-boron nitride heterostructure and photonic crystal nanocavity. *Nano Lett.* **2015**, *15* (3), 2001–2005.
46. Shiue, R.-J.; Gao, Y.; Wang, Y.; Peng, C.; Robertson, A. D.; Efetov, D. K.; Assefa, S.; Koppens, F. H. L.; Hone, J.; Englund, D., High-responsivity graphene–boron nitride photodetector and autocorrelator in a silicon photonic integrated circuit. *Nano Lett.* **2015**, *15* (11), 7288–7293.
47. Lee, G.-H.; Yu, Y.-J.; Cui, X.; Petrone, N.; Lee, C.-H.; Choi, M. S.; Lee, D.-Y.; Lee, C.; Yoo, W. J.; Watanabe, K.; Taniguchi, T.; Nuckolls, C.;

Kim, P.; Hone, J., Flexible and transparent MoS_2 field-effect transistors on hexagonal boron nitride-graphene heterostructures. *ACS Nano.* **2013**, *7* (9), 7931–7936.

48. Withers, F.; Del Pozo-Zamudio, O.; Mishchenko, A.; Rooney, A. P.; Gholinia, A.; Watanabe, K.; Taniguchi, T.; Haigh, S. J.; Geim, A. K.; Tartakovskii, A. I.; Novoselov, K. S., Light-emitting diodes by band-structure engineering in van der Waals heterostructures. *Nat. Mater.* **2015**, *14* (3), 301–306.

49. Britnell, L.; Gorbachev, R. V.; Jalil, R.; Belle, B. D.; Schedin, F.; Mishchenko, A.; Georgiou, T.; Katsnelson, M. I.; Eaves, L.; Morozov, S. V.; Peres, N. M. R.; Leist, J.; Geim, A. K.; Novoselov, K. S.; Ponomarenko, L. A., Field-effect tunneling transistor based on vertical graphene heterostructures. *Science.* **2012**, *335* (6071), 947.

50. Wang, L.; Meric, I.; Huang, P. Y.; Gao, Q.; Gao, Y.; Tran, H.; Taniguchi, T.; Watanabe, K.; Campos, L. M.; Muller, D. A.; Guo, J.; Kim, P.; Hone, J.; Shepard, K. L.; Dean, C. R., One-dimensional electrical contact to a two-dimensional material. *Science.* **2013**, *342* (6158), 614.

51. Mayorov, A. S.; Gorbachev, R. V.; Morozov, S. V.; Britnell, L.; Jalil, R.; Ponomarenko, L. A.; Blake, P.; Novoselov, K. S.; Watanabe, K.; Taniguchi, T.; Geim, A. K., Micrometer-scale ballistic transport in encapsulated graphene at room temperature. *Nano Lett.* **2011**, *11* (6), 2396–2399.

52. Levendorf, M. P.; Kim, C.-J.; Brown, L.; Huang, P. Y.; Havener, R. W.; Muller, D. A.; Park, J., Graphene and boron nitride lateral heterostructures for atomically thin circuitry. *Nature.* **2012**, *488* (7413), 627–632.

53. Liu, Z.; Ma, L.; Shi, G.; Zhou, W.; Gong, Y.; Lei, S.; Yang, X.; Zhang, J.; Yu, J.; Hackenberg, K. P.; Babakhani, A.; Idrobo, J.-C.; Vajtai, R.; Lou, J.; Ajayan, P. M., In-plane heterostructures of graphene and hexagonal boron nitride with controlled domain sizes. *Nat. Nanotech.* **2013**, *8* (2), 119–124.

54. Zhang, C.; Zhao, S.; Jin, C.; Koh, A. L.; Zhou, Y.; Xu, W.; Li, Q.; Xiong, Q.; Peng, H.; Liu, Z., Direct growth of large-area graphene and boron nitride heterostructures by a co-segregation method. *Nature Communications.* **2015**, *6* (1), 6519.

55. Zhu, J.; Kang, J.; Kang, J.; Jariwala, D.; Wood, J. D.; Seo, J.-W. T.; Chen, K.-S.; Marks, T. J.; Hersam, M. C., Solution-processed dielectrics based on thickness-sorted two-dimensional hexagonal boron nitride nanosheets. *Nano Lett.* **2015**, *15* (10), 7029–7036.

56. Novoselov, K. S.; Mishchenko, A.; Carvalho, A.; Castro Neto, A. H., 2D materials and van der Waals heterostructures. *Science.* **2016**, *353* (6298), aac9439.

57. Jariwala, D.; Marks, T. J.; Hersam, M. C., Mixed-dimensional van der Waals heterostructures. *Nat. Mater.* **2017**, *16* (2), 170–181.

58. Schulman, D. S.; Arnold, A. J.; Razavieh, A.; Nasr, J.; Das, S., The prospect of two-dimensional heterostructures: A review of recent breakthroughs. *IEEE Nanotechnol. Mag.* **2017**, *11* (2), 6–17.

59. Tan, C.; Cao, X.; Wu, X.-J.; He, Q.; Yang, J.; Zhang, X.; Chen, J.; Zhao, W.; Han, S.; Nam, G.-H.; Sindoro, M.; Zhang, H., Recent advances in ultrathin two-dimensional nanomaterials. *Chem. Rev.* **2017**, *117* (9), 6225–6331.

60. Meng, J.; Wang, D.; Cheng, L.; Gao, M.; Zhang, X., Recent progress in synthesis, properties, and applications of hexagonal boron nitride-based heterostructures. *Nanotechnology.* **2018**, *30* (7), 074003.

61. Geim, A. K.; Grigorieva, I. V., Van der Waals heterostructures. *Nature.* **2013**, *499* (7459), 419–425.

62. Liu, Y.; Weiss, N. O.; Duan, X.; Cheng, H.-C.; Huang, Y.; Duan, X., Van der Waals heterostructures and devices. *Nat. Rev. Mater.* **2016**, *1* (9), 16042.

63. Liu, Y.; Huang, Y.; Duan, X., Van der Waals integration before and beyond two-dimensional materials. *Nature.* **2019**, *567* (7748), 323–333.

64. Zhang, K.; Feng, Y.; Wang, F.; Yang, Z.; Wang, J., Two dimensional hexagonal boron nitride (2D-h-BN): Synthesis, properties and applications. *J. Mater. Chem. C* **2017**, *5* (46), 11992–12022.

65. Sławińska, J.; Zasada, I.; Klusek, Z., Energy gap tuning in graphene on hexagonal boron nitride bilayer system. *Phys. Rev. B.* **2010**, *81* (15), 155433.

66. Fan, Y.; Zhao, M.; Wang, Z.; Zhang, X.; Zhang, H., Tunable electronic structures of graphene/boron nitride heterobilayers. *Appl. Phys. Lett.* **2011**, *98* (8), 083103.

67. Jung, J.; DaSilva, A. M.; MacDonald, A. H.; Adam, S., Origin of band gaps in graphene on hexagonal boron nitride. *Nat. Commun.* **2015**, *6* (1), 6308.

68. Ju, L.; Velasco, J.; Huang, E.; Kahn, S.; Nosiglia, C.; Tsai, H.-Z.; Yang, W.; Taniguchi, T.; Watanabe, K.; Zhang, Y.; Zhang, G.; Crommie, M.; Zettl, A.; Wang, F., Photoinduced doping in heterostructures of graphene and boron nitride. *Nat. Nanotech.* **2014**, *9* (5), 348–352.

69. Watanabe, K.; Taniguchi, T., Hexagonal boron nitride as a new ultraviolet luminescent material and its application. *International Journal of Applied Ceramic Technology.* **2011**, *8* (5), 977–989.

70. Watanabe, K.; Taniguchi, T.; Kanda, H., Direct-bandgap properties and evidence for ultraviolet lasing of hexagonal boron nitride single crystal. *Nat. Mater.* **2004**, *3* (6), 404–409.

71. Choi, I.; Lee, K.; Lee, C.-R.; Lee, J. S.; Kim, S. M.; Jeong, K.-U.; Kim, J. S., Application of hexagonal boron nitride to a heat-transfer medium of an InGaN/GaN quantum-well green LED. *ACS Appl. Mater. Int.* **2019**, *11* (20), 18876–18884.

72. Watanabe, K.; Taniguchi, T.; Niiyama, T.; Miya, K.; Taniguchi, M., Far-ultraviolet plane-emission handheld device based on hexagonal boron nitride. *Nat. Photonics.* **2009**, *3* (10), 591–594.

73. Kubota, Y.; Watanabe, K.; Tsuda, O.; Taniguchi, T., Deep ultraviolet light-emitting hexagonal boron nitride synthesized at atmospheric pressure. *Science.* **2007**, *317* (5840), 932.

74. Caldwell, J. D.; Aharonovich, I.; Cassabois, G.; Edgar, J. H.; Gil, B.; Basov, D. N., Photonics with hexagonal boron nitride. *Nat. Rev. Mater.* **2019**, *4* (8), 552–567.

75. Ghobadi, N.; Pourfath, M., Vertical tunneling graphene heterostructure-based transistor for pressure sensing. *IEEE Electron Device Lett.* **2015**, *36* (3), 280–282.

76. Xu, Y.; Guo, Z.; Chen, H.; Yuan, Y.; Lou, J.; Lin, X.; Gao, H.; Chen, H.; Yu, B., In-plane and tunneling pressure sensors based on graphene/hexagonal boron nitride heterostructures. *Appl. Phys. Lett.* **2011**, *99* (13), 133109.

77. Li, M.; Zhang, T.; Wang, P.; Li, M.; Wang, J.; Liu, Z., Temperature characteristics of a pressure sensor based on BN/graphene/BN heterostructure. *Sensors.* **2019**, *19* (10).

78. Dankert, A.; Karpiak, B.; Dash, S. P., Hall sensors batch-fabricated on all-CVD h-BN/graphene/h-BN heterostructures. *Sci. Rep.* **2017**, *7* (1), 15231.

79. Karpiak, B.; Dankert, A.; Dash, S. P., Gate-tunable Hall sensors on large area CVD graphene protected by h-BN with 1D edge contacts. *J. Appl. Phys.* **2017**, *122* (5), 054506.

80. Joo, M.-K.; Kim, J.; Park, J.-H.; Nguyen, V. L.; Kim, K. K.; Lee, Y. H.; Suh, D., Large-scale graphene on hexagonal-bn hall elements: Prediction of sensor performance without magnetic field. *ACS Nano.* **2016**, *10* (9), 8803–8811.

81. Liu, G.; Rumyantsev, S. L.; Jiang, C.; Shur, M. S.; Balandin, A. A., Selective gas sensing with h-BN capped MoS_2 heterostructure thin-film transistors. *IEEE Electron Device Lett.* **2015**, *36* (11), 1202–1204.

82. Gautam, A.; Gautam, C.; Kumar, U.; Yadav, B. C., Synthesis and structural investigations of microporous graphene-reinforced h-BN solids for LPG sensing applications. *Mater. Res. Exp.* **2019**, *6* (12), 125090.

83. de Souza, F. A. L.; Amorim, R. G.; Scopel, W. L.; Scheicher, R. H., Nano-structured interface of graphene and h-BN for sensing applications. *Nanotechnology.* **2016**, *27* (36), 365503.

84. Wei, W.; Liao, W.; Zeng, Z.; Zhu, C., Extended gate reference-FET (REFET) using 2D h-BN sensing layer for pH sensing applications. *IEEE Electron Device Lett.* **2020**, *41* (1), 159–162.

85. Sajjad, M.; Feng, P., Study the gas sensing properties of boron nitride nanosheets. *Mater. Res. Bull.* **2014**, *49*, 35–38.

86. Lin, L.; Liu, T.; Zhang, Y.; Sun, R.; Zeng, W.; Wang, Z., Synthesis of boron nitride nanosheets with a few atomic layers and their gas-sensing performance. *Ceram. Int.* **2016**, *42* (1, Part A), 971–975.

87. Weng, Q.; Wang, X.; Wang, X.; Bando, Y.; Golberg, D., Functionalized hexagonal boron nitride nanomaterials: emerging properties and applications. *Chem. Soc. Rev.* **2016**, *45* (14), 3989–4012.

88. Meziani, M. J.; Song, W.-L.; Wang, P.; Lu, F.; Hou, Z.; Anderson, A.; Maimaiti, H.; Sun, Y.-P., Boron nitride nanomaterials for thermal management applications. *ChemPhysChem.* **2015**, *16* (7), 1339–1346.

89. Pakdel, A.; Bando, Y.; Golberg, D., Plasma-assisted interface engineering of boron nitride nanostructure films. *ACS Nano.* **2014**, *8* (10), 10631–10639.

90. Li, T.-L.; Hsu, S. L.-C., Enhanced thermal conductivity of polyimide films *via* a hybrid of micro- and nano-sized boron nitride. *J. Phys. Chem. B.* **2010**, *114* (20), 6825–6829.

91. Li, T.-L.; Hsu, S. L.-C., Preparation and properties of thermally conductive photosensitive polyimide/boron nitride nanocomposites. *J. Appl. Polym. Sci.* **2011**, *121* (2), 916–922.

92. Shoji, Y.; Higashihara, T.; Tokita, M.; Morikawa, J.; Watanabe, J.; Ueda, M., Thermal diffusivity of hexagonal boron nitride composites based on cross-linked liquid crystalline polyimides. *ACS Appl. Mater. Int.* **2013**, *5* (8), 3417–3423.

93. Yoshihara, S.; Sakaguchi, M.; Matsumoto, K.; Tokita, M.; Watanabe, J., Influence of molecular orientation direction on the in-plane thermal conductivity of polymer/hexagonal boron nitride composites. *J. Appl. Polym. Sci.* **2014**, *131* (3).

94. Wang, X.; Pakdel, A.; Zhang, J.; Weng, Q.; Zhai, T.; Zhi, C.; Golberg, D.; Bando, Y., Large-surface-area BN nanosheets and their utilization in polymeric composites with improved thermal and dielectric properties. *Nanoscale Res. Lett.* **2012**, *7* (1), 662.

95. Wang, Z.; Fu, Y.; Meng, W.; Zhi, C., Solvent-free fabrication of thermally conductive insulating epoxy composites with boron nitride nanoplatelets as fillers. *Nanoscale Res. Lett.* **2014**, *9* (1), 643.

96. Lin, Z.; McNamara, A.; Liu, Y.; Moon, K.-S.; Wong, C.-P., Exfoliated hexagonal boron nitride-based polymer nanocomposite with enhanced thermal conductivity for electronic encapsulation. *Compos. Sci. Technol.* **2014**, *90*, 123–128.

97. Wang, X.-B.; Weng, Q.; Wang, X.; Li, X.; Zhang, J.; Liu, F.; Jiang, X.-F.; Guo, H.; Xu, N.; Golberg, D.; Bando, Y., Biomass-directed synthesis of 20 g high-quality boron nitride nanosheets for thermoconductive polymeric composites. *ACS Nano.* **2014**, *8* (9), 9081–9088.

98. Li, S.; Yang, T.; Zou, H.; Liang, M.; Chen, Y., Enhancement in thermal conductivity and mechanical properties *via* large-scale fabrication of boron nitride nanosheets. *High Perform. Polym.* **2016**, *29* (3), 315–327.

99. Xie, S.; Istrate, O. M.; May, P.; Barwich, S.; Bell, A. P.; Khan, U.; Coleman, J. N., Boron nitride nanosheets as barrier enhancing fillers in melt processed composites. *Nanoscale.* **2015**, *7* (10), 4443–4450.

100. Wu, X.; Liu, H.; Tang, Z.; Guo, B., Scalable fabrication of thermally conductive elastomer/boron nitride nanosheets composites by slurry compounding. *Compos. Sci. Technol.* **2016**, *123*, 179–186.

101. Cho, H.-B.; Mitsuhashi, M.; Nakayama, T.; Tanaka, S.; Suzuki, T.; Suematsu, H.; Jiang, W.; Tokoi, Y.; Lee, S. W.; Park, Y.-H.; Niihara, K., Thermal anisotropy of epoxy resin-based nano-hybrid films containing BN nanosheets under a rotating superconducting magnetic field. *Mater. Chem. Phys.* **2013**, *139* (2), 355–359.

102. Yu, J.; Huang, X.; Wu, C.; Wu, X.; Wang, G.; Jiang, P., Interfacial modification of boron nitride nanoplatelets for epoxy composites with improved thermal properties. *Polymer.* **2012**, *53* (2), 471–480.

103. Song, W.-L.; Wang, P.; Cao, L.; Anderson, A.; Meziani, M. J.; Farr, A. J.; Sun, Y.-P., Polymer/boron nitride nanocomposite materials for superior thermal transport performance. *Angew. Chem. Int. Ed.* **2012**, *51* (26), 6498–6501.

104. Taha-Tijerina, J.; Narayanan, T. N.; Gao, G.; Rohde, M.; Tsentalovich, D. A.; Pasquali, M.; Ajayan, P. M., Electrically insulating thermal nano-oils using 2D fillers. *ACS Nano.* **2012**, *6* (2), 1214–1220.

105. Zhu, H.; Li, Y.; Fang, Z.; Xu, J.; Cao, F.; Wan, J.; Preston, C.; Yang, B.; Hu, L., Highly thermally conductive papers with percolative layered boron nitride nanosheets. *ACS Nano.* **2014**, *8* (4), 3606–3613.

106. Liu, J.; Kutty, R. G.; Zheng, Q.; Eswariah, V.; Sreejith, S.; Liu, Z., Hexagonal boron nitride nanosheets as high-performance binder-free fire-resistant wood coatings. *Small.* **2017**, *13* (2), 1602456.

107. Ambekar, R. S.; Deshmukh, A.; Suárez-Villagrán, M. Y.; Das, R.; Pal, V.; Dey, S.; Miller, J. H.; Machado, L. D.; Kumbhakar, P.; Tiwary, C. S., 2D hexagonal boron nitride-coated cotton fabric with self-extinguishing property. *ACS Appl. Mater. Int.* **2020**, *12* (40), 45274–45280.

108. Boland, C. S.; Barwich, S.; Khan, U.; Coleman, J. N., High stiffness nano-composite fibres from polyvinylalcohol filled with graphene and boron nitride. *Carbon.* **2016**, *99*, 280–288.

109. Jan, R.; May, P.; Bell, A. P.; Habib, A.; Khan, U.; Coleman, J. N., Enhancing the mechanical properties of BN nanosheet–polymer composites by uniaxial drawing. *Nanoscale.* **2014**, *6* (9), 4889–4895.

110. Zhi, C.; Bando, Y.; Tang, C.; Kuwahara, H.; Golberg, D., Large-scale fabrication of boron nitride nanosheets and their utilization in polymeric composites with improved thermal and mechanical properties. *Adv. Mater.* **2009**, *21* (28), 2889–2893.

111. Sainsbury, T.; Satti, A.; May, P.; Wang, Z.; McGovern, I.; Guńko, Y. K.; Coleman, J., Oxygen radical functionalization of boron nitride nanosheets. *J. Am. Chem. Soc.* **2012**, *134* (45), 18758–18771.

112. Saggar, R.; Porwal, H.; Tatarko, P.; Dlouhý, I.; Reece, M. J., Boron nitride nanosheets reinforced glass matrix composites. *Adv. Appl. Ceram.* **2015**, *114* (sup1), S26–S33.

113. Liu, Z.; Gong, Y.; Zhou, W.; Ma, L.; Yu, J.; Idrobo, J. C.; Jung, J.; MacDonald, A. H.; Vajtai, R.; Lou, J.; Ajayan, P. M., Ultrathin high-temperature oxidation-resistant coatings of hexagonal boron nitride. *Nature Commun.* **2013**, *4*, 2541.

114. Khan, M. H.; Jamali, S. S.; Lyalin, A.; Molino, P. J.; Jiang, L.; Liu, H. K.; Taketsugu, T.; Huang, Z., Atomically thin hexagonal boron nitride nanofilm for Cu protection: The importance of film perfection. *Adv. Mater.* **2017**, *29* (4), 1603937.

115. Mahvash, F.; Eissa, S.; Bordjiba, T.; Tavares, A. C.; Szkopek, T.; Siaj, M., Corrosion resistance of monolayer hexagonal boron nitride on copper. *Sci. Rep.* **2017**, *7* (1), 42139.

116. Chilkoor, G.; Jawaharraj, K.; Vemuri, B.; Kutana, A.; Tripathi, M.; Kota, D.; Arif, T.; Filleter, T.; Dalton, A. B.; Yakobson, B. I.; Meyyappan,

M.; Rahman, M. M.; Ajayan, P. M.; Gadhamshetty, V., Hexagonal boron nitride for sulfur corrosion inhibition. *ACS Nano.* **2020**, *14* (11), 14809–14819.

117. Chilkoor, G.; Karanam, S. P.; Star, S.; Shrestha, N.; Sani, R. K.; Upadhyayula, V. K. K.; Ghoshal, D.; Koratkar, N. A.; Meyyappan, M.; Gadhamshetty, V., Hexagonal boron nitride: The thinnest insulating barrier to microbial corrosion. *ACS Nano.* **2018**, *12* (3), 2242–2252.

118. Cui, M.; Ren, S.; Chen, J.; Liu, S.; Zhang, G.; Zhao, H.; Wang, L.; Xue, Q., Anticorrosive performance of waterborne epoxy coatings containing water-dispersible hexagonal boron nitride (h-BN) nanosheets. *Appl. Surf. Sci.* **2017**, *397*, 77–86.

119. Cui, M.; Ren, S.; Qin, S.; Xue, Q.; Zhao, H.; Wang, L., Non-covalent functionalized hexagonal boron nitride nanoplatelets to improve corrosion and wear resistance of epoxy coatings. *RSC Adv.* **2017**, *7* (70), 44043–44053.

120. Cui, M.; Ren, S.; Qin, S.; Xue, Q.; Zhao, H.; Wang, L., Processable poly(2-butylaniline)/hexagonal boron nitride nanohybrids for synergetic anticorrosive reinforcement of epoxy coating. *Corros. Sci.* **2018**, *131*, 187–198.

121. Martinez Rubi, Y.; Jakubek, Z. J.; Chen, M.; Zou, S.; Simard, B., Quality assessment of bulk boron nitride nanotubes for advancing research, commercial, and industrial applications. *ACS. Appl. Nano Mater.* **2019**, *2* (4), 2054–2063.

122. Zhi, C.; Bando, Y.; Tang, C.; Honda, S.; Kuwahara, H.; Golberg, D., Boron nitride nanotubes/polystyrene composites. *J. Mater. Res.* **2006**, *21* (11), 2794–2800.

123. Zhi, C.; Bando, Y.; Terao, T.; Tang, C.; Kuwahara, H.; Golberg, D., Towards thermoconductive, electrically insulating polymeric composites with boron nitride nanotubes as fillers. *Adv. Funct. Mater.* **2009**, *19* (12), 1857–1862.

124. Zhi, C. Y.; Bando, Y.; Terao, T.; Tang, C. C.; Kuwahara, H.; Golberg, D., Chemically activated boron nitride nanotubes. *Chem. Asian J.* **2009**, *4* (10), 1536–1540.

125. Zhou, S.-J.; Ma, C.-Y.; Meng, Y.-Y.; Su, H.-F.; Zhu, Z.; Deng, S.-L.; Xie, S.-Y., Activation of boron nitride nanotubes and their polymer composites for improving mechanical performance. *Nanotechnology.* **2012**, *23* (5), 055708.

126. Yan, H.; Tang, Y.; Su, J.; Yang, X., Enhanced thermal–mechanical properties of polymer composites with hybrid boron nitride nanofillers. *Appl. Phys. A* **2014**, *114* (2), 331–337.

127. Lu, X.; Nautiyal, P.; Bustillos, J.; Loganathan, A.; Zhang, C.; Chen, Y.; Boesl, B.; Agarwal, A., Hydroxylated boron nitride nanotube-reinforced polyvinyl alcohol nanocomposite films with simultaneous improvement of mechanical and thermal properties. *Polym. Compos.* **2020**, *41* (12), 5182–5194.

128. Kim, K. S.; Jakubinek, M. B.; Martinez-Rubi, Y.; Ashrafi, B.; Guan, J.; O'Neill, K.; Plunkett, M.; Hrdina, A.; Lin, S.; Dénommée, S.; Kingston, C.; Simard, B., Polymer nanocomposites from free-standing, macroscopic boron nitride nanotube assemblies. *RSC Adv.* **2015**, *5* (51), 41186–41192.

129. Yamaguchi, M.; Pakdel, A.; Zhi, C.; Bando, Y.; Tang, D.-M.; Faerstein, K.; Shtansky, D.; Golberg, D., Utilization of multiwalled boron nitride nanotubes for the reinforcement of lightweight aluminum ribbons. *Nanoscale Res. Lett.* **2013**, *8* (1), 3.

130. Azizi, K.; Salabat, K.; Seif, A., Methane storage on aluminum-doped single wall BNNTs. *Appl. Surf. Sci.* **2014**, *309*, 54–61.

131. Wang, M.; Zhang, T.; Mao, D.; Yao, Y.; Zeng, X.; Ren, L.; Cai, Q.; Mateti, S.; Li, L. H.; Zeng, X.; Du, G.; Sun, R.; Chen, Y.; Xu, J.-B.; Wong, C.-P., Highly compressive boron nitride nanotube aerogels reinforced with reduced graphene oxide. *ACS Nano.* **2019**, *13* (7), 7402–7409.

132. Behabtu, N.; Young, C. C.; Tsentalovich, D. E.; Kleinerman, O.; Wang, X.; Ma, A. W. K.; Bengio, E. A.; ter Waarbeek, R. F.; de Jong, J. J.; Hoogerwerf, R. E.; Fairchild, S. B.; Ferguson, J. B.; Maruyama, B.; Kono, J.; Talmon, Y.; Cohen, Y.; Otto, M. J.; Pasquali, M., Strong, light, multifunctional fibers of carbon nanotubes with ultrahigh conductivity. *Science.* **2013**, *339* (6116), 182.

133. Ericson, L. M.; Fan, H.; Peng, H.; Davis, V. A.; Zhou, W.; Sulpizio, J.; Wang, Y.; Booker, R.; Vavro, J.; Guthy, C.; Parra-Vasquez, A. N. G.; Kim, M. J.; Ramesh, S.; Saini, R. K.; Kittrell, C.; Lavin, G.; Schmidt, H.; Adams, W. W.; Billups, W. E.; Pasquali, M.; Hwang, W.-F.; Hauge, R. H.; Fischer, J. E.; Smalley, R. E., Macroscopic, neat, single-walled carbon nanotube fibers. *Science.* **2004**, *305* (5689), 1447.

134. Li, Y.-L.; Kinloch, I. A.; Windle, A. H., Direct spinning of carbon nanotube fibers from chemical vapor deposition synthesis. *Science.* **2004**, *304* (5668), 276.

135. Zhang, M.; Atkinson, K. R.; Baughman, R. H., Multifunctional carbon nanotube yarns by downsizing an ancient technology. *Science.* **2004**, *306* (5700), 1358.

136. Smith, M. W.; Jordan, K. C.; Park, C.; Kim, J.-W.; Lillehei, P. T.; Crooks, R.; Harrison, J. S., Very long single- and few-walled boron nitride nanotubes *via* the pressurized vapor/condenser method. *Nanotechnology.* **2009**, *20* (50), 505604.

137. Kim, K. S.; Kingston, C. T.; Hrdina, A.; Jakubinek, M. B.; Guan, J.; Plunkett, M.; Simard, B., Hydrogen-catalyzed, pilot-scale production of small-diameter boron nitride nanotubes and their macroscopic assemblies. *ACS Nano.* **2014**, *8* (6), 6211–6220.

138. Fathalizadeh, A.; Pham, T.; Mickelson, W.; Zettl, A., Scaled synthesis of boron nitride nanotubes, nanoribbons, and nanococoons using direct feedstock injection into an extended-pressure, inductively-coupled thermal plasma. *Nano Lett.* **2014**, *14* (8), 4881–4886.

139. Huang, X.; Zhi, C.; Jiang, P.; Golberg, D.; Bando, Y.; Tanaka, T., Polyhedral oligosilsesquioxane-modified boron nitride nanotube based epoxy nanocomposites: An ideal dielectric material with high thermal conductivity. *Adv. Funct. Mater.* **2013**, *23* (14), 1824–1831.

140. Jang, I.; Shin, K.-H.; Yang, I.; Kim, H.; Kim, J.; Kim, W.-H.; Jeon, S.-W.; Kim, J.-P., Enhancement of thermal conductivity of BN/epoxy composite through surface modification with silane coupling agents. *Colloids Surf. A Physicochem. Eng.* **2017**, *518*, 64–72.

141. Fu, C.; Li, Q.; Lu, J.; Mateti, S.; Cai, Q.; Zeng, X.; Du, G.; Sun, R.; Chen, Y.; Xu, J.; Wong, C.-P., Improving thermal conductivity of polymer composites by reducing interfacial thermal resistance between boron nitride nanotubes. *Compos. Sci. Technol.* **2018**, *165*, 322–330.

142. Quiles-Díaz, S.; Martínez-Rubí, Y.; Guan, J.; Kim, K. S.; Couillard, M.; Salavagione, H. J.; Gómez-Fatou, M. A.; Simard, B., Enhanced thermal conductivity in polymer nanocomposites *via* covalent functionalization of boron nitride nanotubes with short polyethylene chains for heat-transfer applications. *ACS. Appl. Nano Mater.* **2019**, *2* (1), 440–451.

143. Guan, J.; Derdouri, A.; Ashrafi, B.; Benhalima, A.; Kim, K. S.; Daroszewska, M.; Simard, B., Boron nitride nanotubes reinforced polycarbonate nanocomposites. *Mater. Today Commun.* **2019**, *20*, 100586.

144. Wang, L.; Han, D.; Luo, J.; Li, T.; Lin, Z.; Yao, Y., Highly efficient growth of boron nitride nanotubes and the thermal conductivity of their polymer composites. *J. Phys. Chem. C.* **2018**, *122* (3), 1867–1873.

145. Zeng, X.; Sun, J.; Yao, Y.; Sun, R.; Xu, J.-B.; Wong, C.-P., A combination of boron nitride nanotubes and cellulose nanofibers for the preparation of a nanocomposite with high thermal conductivity. *ACS Nano.* **2017**, *11* (5), 5167–5178.

146. Zhi, C.; Xu, Y.; Bando, Y.; Golberg, D., Highly thermo-conductive fluid with boron nitride nanofillers. *ACS Nano.* **2011**, *5* (8), 6571–6577.

147. Seo, T. H.; Lee, G. H.; Park, A. H.; Cho, H.; Kim, J.-H.; Chandramohan, S.; Jeon, S.-R.; Jang, S. G.; Kim, M. J.; Suh, E.-K., Boron nitride nanotubes as a heat sinking and stress-relaxation layer for high performance light-emitting diodes. *Nanoscale.* **2017**, *9* (42), 16223–16231.

148. Li, L. L.; Yang, S. Q.; Yang, X. J.; Xu, X. W.; Tang, C. C., Boron adsorption induced magnetism in zigzag boron nitride nanotubes. *J. Mol. Struct.* **2012**, *1020*, 183–187.

149. Wu, R. Q.; Liu, L.; Peng, G. W.; Feng, Y. P., Magnetism in BN nanotubes induced by carbon doping. *Appl. Phys. Lett.* **2005**, *86* (12), 122510.

150. Moradian, R.; Azadi, S., Magnetism in defected single-walled boron nitride nanotubes. *EPL.* **2008**, *83* (1), 17007.

151. Wei, X.; Wang, M.-S.; Bando, Y.; Golberg, D., Electron-beam-induced substitutional carbon doping of boron nitride nanosheets, nanoribbons, and nanotubes. *ACS Nano.* **2011**, *5* (4), 2916–2922.

152. Xiang, H. J.; Liang, W. Z.; Yang, J.; Hou, J. G.; Zhu, Q., Spin-unrestricted linear-scaling electronic structure theory and its application to magnetic carbon-doped boron nitride nanotubes. *J. Chem. Phys.* **2005**, *123* (12), 124105.

153. Zhang, Z.; Guo, W., Tunable ferromagnetic spin ordering in boron nitride nanotubes with topological fluorine adsorption. *J. Am. Chem. Soc.* **2009**, *131* (19), 6874–6879.

154. Li, F.; Zhu, Z.; Yao, X.; Lu, G.; Zhao, M.; Xia, Y.; Chen, Y., Fluorination-induced magnetism in boron nitride nanotubes from *ab initio* calculations. *Appl. Phys. Lett.* **2008**, *92* (10), 102515.

155. Si, M. S.; Xue, D. S., First-principles study of silicon-doped (5,5) BN nanotubes. *Europhysics Letters (EPL).* **2006**, *76* (4), 664–669.

156. Wu, J.; Zhang, W., Magnetism in germanium-doped boron-nitride nanotubes. *Chem. Phys. Lett.* **2008**, *457* (1), 169–173.

157. Zhang, J. M.; Wang, S. F.; Chen, L. Y.; Xu, K. W.; Ji, V., Structural, electronic and magnetic properties of the 3D transition metal atoms adsorbed on boron nitride nanotubes. *Eur. Phys. J. B.* **2010**, *76* (2), 289–299.

158. Wu, X.; Zeng, X. C., Adsorption of transition-metal atoms on boron nitride nanotube: A density-functional study. *J. Chem. Phys.* **2006**, *125* (4), 044711.

159. Dhungana, K. B.; Pati, R., Boron nitride nanotubes for spintronics. *Sensors.* **2014**, *14* (9).

160. Hao, S.; Zhou, G.; Duan, W.; Wu, J.; Gu, B.-L., Tremendous spin-splitting effects in open boron nitride nanotubes: Application to nanoscale spintronic devices. *J. Am. Chem. Soc.* **2006**, *128* (26), 8453–8458.

161. Li, J.; Zhou, G.; Chen, Y.; Gu, B.-L.; Duan, W., Magnetism of C adatoms on BN nanostructures: Implications for functional nanodevices. *J. Am. Chem. Soc.* **2009**, *131* (5), 1796–1801.

162. Lenin, A.; Arumugam, P.; Shanmugham, R.; Sonachalam, A.; Paramasivam, S.; Rao, A. P.; Singaravelu, G.; Venkatesan, R., Hybrid Ni–boron nitride nanotube magnetic semiconductor—A new material for spintronics. *ACS Omega* **2020**, *5* (32), 20014–20020.

163. Sengupta, A., Atomistic study of electrostatics and carrier transport properties of CNT@MS_2 (M = Mo, W) and CNT@BN core–shell nanotubes. *J. Mater. Sci.* **2017**, *52* (13), 8119–8131.

164. Hu, C.; Michaud-Rioux, V.; Yao, W.; Guo, H., Theoretical design of topological heteronanotubes. *Nano Lett.* **2019**, *19* (6), 4146–4150.

165. An, Y.; Sun, Y.; Jiao, J.; Zhang, M.; Wang, K.; Chen, X.; Wu, D.; Wang, T.; Fu, Z.; Jiao, Z., The rectifying effect of heterojunctions composed of carbon and boron nitride nanotubes. *Org. Electron.* **2017**, *50*, 43–47.

166. Berdiyorov, G. R.; Hamoudi, H., Doping-enhanced current rectification in carbon nanotube: Metal junctions for rectenna applications. *ACS Omega.* **2020**, *5* (1), 189–196.

167. Chen, X.-K.; Xie, Z.-X.; Zhang, Y.; Deng, Y.-X.; Zou, T.-H.; Liu, J.; Chen, K.-Q., Highly efficient thermal rectification in carbon/boron nitride heteronanotubes. *Carbon.* **2019**, *148*, 532–539.

168. Vedaei, S. S.; Nadimi, E., Gas sensing properties of CNT-BNNT-CNT nanostructures: A first principles study. *Appl. Surf. Sci.* **2019**, *470*, 933–942.

169. Xiang, R.; Inoue, T.; Zheng, Y.; Kumamoto, A.; Qian, Y.; Sato, Y.; Liu, M.; Tang, D.; Gokhale, D.; Guo, J.; Hisama, K.; Yotsumoto, S.; Ogamoto, T.; Arai, H.; Kobayashi, Y.; Zhang, H.; Hou, B.; Anisimov, A.; Maruyama, M.; Miyata, Y.; Okada, S.; Chiashi, S.; Li, Y.; Kong, J.; Kauppinen, E. I.; Ikuhara, Y.; Suenaga, K.; Maruyama, S., One-dimensional van der Waals heterostructures. *Science.* **2020**, *367* (6477), 537–542.

170. Kang, J. H.; Sauti, G.; Park, C.; Yamakov, V. I.; Wise, K. E.; Lowther, S. E.; Fay, C. C.; Thibeault, S. A.; Bryant, R. G., Multifunctional electroactive nanocomposites based on piezoelectric boron nitride nanotubes. *ACS Nano.* **2015**, *9* (12), 11942–11950.

171. Lahiri, D.; Singh, V.; Benaduce, A. P.; Seal, S.; Kos, L.; Agarwal, A., Boron nitride nanotube reinforced hydroxyapatite composite: Mechanical and tribological performance and *in vitro* biocompatibility to osteoblasts. *J Mech. Behav. Biomed. Mater.* **2011**, *4* (1), 44–56.

172. Ciofani, G.; Ricotti, L.; Danti, S.; Moscato, S.; Nesti, C.; D'Alessandro, D.; Dinucci, D.; Chiellini, F.; Pietrabissa, A.; Petrini, M.; Menciassi, A., Investigation of interactions between poly-L-lysine-coated boron nitride nanotubes and C_2C_{12} cells: Up-take, cytocompatibility, and differentiation. *Int. J. Nanomedicine.* **2010**, *5*, 285–298.

173. Lahiri, D.; Rouzaud, F.; Richard, T.; Keshri, A. K.; Bakshi, S. R.; Kos, L.; Agarwal, A., Boron nitride nanotube reinforced polylactide–polycaprolactone copolymer composite: Mechanical properties and cytocompatibility with osteoblasts and macrophages in vitro. *Acta Biomater.* **2010**, *6* (9), 3524–3533.

174. Gao, C.; Feng, P.; Peng, S.; Shuai, C., Carbon nanotube, graphene and boron nitride nanotube reinforced bioactive ceramics for bone repair. *Acta Biomater.* **2017**, *61*, 1–20.

175. Kalay, S. Y., Z.; Sen, O.; Emanet, M.; Kazanc, E.; Çulha, M., Synthesis of boron nitride nanotubes and their applications. *Beilstein J. Nanotechnol.* **2015**, *6*, 84–102.

176. Chen, X.; Wu, P.; Rousseas, M.; Okawa, D.; Gartner, Z.; Zettl, A.; Bertozzi, C. R., Boron nitride nanotubes are noncytotoxic and can be functionalized for interaction with proteins and cells. *J. Am. Chem. Soc.* **2009**, *131* (3), 890–891.

177. Horváth, L.; Magrez, A.; Golberg, D.; Zhi, C.; Bando, Y.; Smajda, R.; Horváth, E.; Forró, L.; Schwaller, B., *in vitro* investigation of the cellular toxicity of boron nitride nanotubes. *ACS Nano.* **2011**, *5* (5), 3800–3810.

178. Del Turco, S.; Ciofani, G.; Cappello, V.; Gemmi, M.; Cervelli, T.; Saponaro, C.; Nitti, S.; Mazzolai, B.; Basta, G.; Mattoli, V., Cytocompatibility evaluation of glycol-chitosan coated boron nitride nanotubes in human endothelial cells. *Colloids Surf. B.* **2013**, *111*, 142–149.

179. G., C.; L., R.; S., D.; Moscato, S.; Nesti C; D'Alessandro.; D., D.; F., C.; A., P.; M., P.; A., M., Investigation of interactions between poly-L-lysine-coated boron nitride nanotubes and C2C12 cells: Up-take, cytocompatibility, and differentiation. *Int. J. Nanomedicine.* **2010**, *5*, 285–298.

180. Ciofani, G.; Raffa, V.; Menciassi, A.; Cuschieri, A., Folate functionalized boron nitride nanotubes and their selective uptake by glioblastoma

multiforme cells: Implications for their use as boron carriers in clinical boron neutron capture therapy. *Nanoscale Res. Lett.* **2008**, *4* (2), 113.

181. Ciofani, G.; Boni, A.; Calucci, L.; Forte, C.; Gozzi, A.; Mazzolai, B.; Mattoli, V., Gd-doped BNNTs as T_2-weighted MRI contrast agents. *Nanotechnology.* **2013**, *24* (31), 315101.

182. Ferreira, T. H.; Soares, D. C. F.; Moreira, L. M. C.; da Silva, P. R. O.; dos Santos, R. G.; de Sousa, E. M. B., Boron nitride nanotubes coated with organic hydrophilic agents: Stability and cytocompatibility studies. *Mater. Sci. Eng. C.* **2013**, *33* (8), 4616–4623.

183. G., C.; S., D.; Graziana, G. G.; D., D.; J., P.; M., O.; V., M.; M., G., Pilot in vivo toxicological investigation of boron nitride nanotubes. *Int. J. Nanomedicine.* **2012**, *7*, 19–24.

184. Ciofani, G.; Danti, S.; Nitti, S.; Mazzolai, B.; Mattoli, V.; Giorgi, M., Biocompatibility of boron nitride nanotubes: An up-date of in vivo toxicological investigation. *Int. J. Pharm.* **2013**, *444* (1), 85–88.

185. Soares, D. C. F.; Ferreira, T. H.; Ferreira, C. d. A.; Cardoso, V. N.; de Sousa, E. M. B., Boron nitride nanotubes radiolabeled with ^{99}mTc: Preparation, physicochemical characterization, biodistribution study, and scintigraphic imaging in Swiss mice. *Int. J. Pharm.* **2012**, *423* (2), 489–495.

186. Li, X.; Zhi, C.; Hanagata, N.; Yamaguchi, M.; Bando, Y.; Golberg, D., Boron nitride nanotubes functionalized with mesoporous silica for intracellular delivery of chemotherapy drugs. *Chem. Commun.* **2013**, *49* (66), 7337–7339.

187. Ciofani, G. R.; Yu, J.; Chen, Y. O.; Takeoka, S.; Menciassi, A.; Cuschieri, A., Boron nitride nanotubes: A novel vector for targeted magnetic drug delivery. *Current Nanoscience.* **2009**, *5*, 33–38.

188. Li, X.; Hanagata, N.; Wang, X.; Yamaguchi, M.; Yi, W.; Bando, Y.; Golberg, D., Multimodal luminescent-magnetic boron nitride nanotubes@ NaGdF4: Eu structures for cancer therapy. *Chem. Commun.* **2014**, *50* (33), 4371–4374.

189. Shakerzadeh, E.; Noorizadeh, S., A first principles study of pristine and Al-doped boron nitride nanotubes interacting with platinum-based anticancer drugs. *Physica E.* **2014**, *57*, 47–55.

190. Cho, H.; Kim, J. H.; Hwang, J. H.; Kim, C. S.; Jang, S. G.; Park, C.; Lee, H.; Kim, M. J., Single- and double-walled boron nitride nanotubes: Controlled synthesis and application for water purification. *Sci. Rep.* **2020**, *10* (1), 7416.

191. Huang, Q.; Bando, Y.; Zhao, L.; Zhi, C. Y.; Golberg, D., pH sensor based on boron nitride nanotubes. *Nanotechnology.* **2009**, *20* (41), 415501.

192. Yu, Y.; Chen, H.; Liu, Y.; Li, L. H.; Chen, Y., Humidity sensing properties of single Au-decorated boron nitride nanotubes. *Electrochem Commun.* **2013**, *30*, 29–33.

193. Kim, I.; Roh, H.; Yu, J.; Jayababu, N.; Kim, D., Boron nitride nanotube-based contact electrification-assisted piezoelectric nanogenerator as a kinematic sensor for detecting the flexion–extension motion of a robot finger. *ACS. Energy Lett.* **2020**, *5* (5), 1577–1585.

194. Won, C. Y.; Aluru, N. R., Water permeation through a subnanometer boron nitride nanotube. *J. Am. Chem. Soc.* **2007**, *129* (10), 2748–2749.

Acronym List

AAO:	anodic aluminum oxide
AFM:	atomic force microscopy
BN:	boron nitride
BNNS:	boron nitride nanosheet
BNNT:	boron nitride nanotube
BOCVD:	boron oxide chemical vapor deposition
c-BN:	cubic boron nitride
CNT:	carbon nanotube
CVD:	chemical vapor deposition
DC:	direct current
DCM:	dichloromethane
DFT:	density functional theory
DFTB:	density functional tight binding
DMSO:	dimethyl sulfoxide
DNA:	deoxyribonucleic acid
ECR:	electron cyclotron resonance
EELS:	electron energy loss spectroscopy
FET:	field effect transistor
FME:	flow rate modulated epitaxy
h-BN:	hexagonal boron nitride
HRSTM:	high resolution scanning tunneling microscope
HRTEM:	high resolution transmission electron microscope
IPA:	isopropyl alcohol
IR:	infrared
LDA:	local density approximation
LEED:	low-energy electron diffraction
LO:	longitudinal optical
MBE:	molecular beam epitaxy
MD:	molecular dynamics
MO:	mineral oil
MWBNNT:	multi-walled boron nitride nanotube
MWCNT:	multi-walled carbon nanotube
N-ZZ:	N-terminated zigzag

PcBN:	polycrystalline cubic boron nitride
PcD:	polycrystalline diamond
PDDA:	polydiallyldimethylammonium chloride
PEVA:	poly(ethylene vinyl alcohol)
PL:	photoluminescence
PLD:	pulsed laser deposition
PMMA:	polymethyl methacrylate
POSS:	polyhedral oligosilsesquioxane
PS:	polystyrene
PVA:	polyvinyl alcohol
PVB:	poly(vinyl butyral)
r-BN:	rhombohedral boron nitride
RF:	radio frequency
rGO:	reduced graphene oxide
SEM:	scanning electron microscope
SLS:	solid-liquid-solid
STM:	scanning tunneling microscope
SWBNNT:	single-walled boron nitride nanotube
SWCNT:	single-walled carbon nanotube
TEM:	transmission electron microscope
TGA:	thermogravimetric analysis
TMD:	transition metal dichalcogenide
TO:	transverse optical
TOF-SIMS:	time-of-flight second ion mass spectrometry
TPU:	thermopolyurethane
VLS:	vapor-liquid-solid
w-BN:	wurtzite boron nitride
XPD:	X-ray photoelectron diffraction
XPS:	X-ray photoelectron spectroscopy
XRD:	X-ray diffraction
ZZ:	zigzag

Index

For Product Safety Concerns and Information please contact our EU representative GPSR@taylorandfrancis.com Taylor & Francis Verlag GmbH, Kaufingerstraße 24, 80331 München, Germany